Pest Control for Professional Turfgrass Managers

2022

This publication is also available on the NC State Extension Publications Catalog at
content.ces.ncsu.edu/pest-control-for-professional-turfgrass-managers

Published by

NC State Extension
North Carolina State University

Table of Contents

Recommendations of specific chemicals are based on information on the manufacturer's label and performance in a limited number of trials. Because environmental conditions and methods of application by growers may vary widely, performance of the chemical will not always conform to the safety and pest control standards indicated by experimental data.

Recommendations for the use of agricultural chemicals are included in this publication as a convenience to the reader. The use of brand names and any mention or listing of commercial products or services in this publication does not imply endorsement by NC State Extension nor discrimination against similar products or services not mentioned. Individuals who use agricultural chemicals are responsible for ensuring that the intended use complies with current regulations and conforms to the product label. Be sure to obtain current information about usage regulations and examine a current product label before applying any chemical. For assistance, contact your county Cooperative Extension agent.

Introduction

Pest Control for Professional Turfgrass Managers supplies up-to-date information on pesticides used to control pests in turfgrasses. The tables in this book supplement information available in other NC State Extension publications. The *Turfgrass Pest Management Manual,* AG-348, provides information that will help the reader identify major turfgrasses and turfgrass pests and better understand their life cycles, symptoms, and biology.

Certain pesticides may be used only by commercial or professional and landscape personnel. "GENERAL USE" pesticides can be purchased and applied by anyone; however, anyone applying any pesticide for pay or to public property (including golf courses) must have a license or be supervised by someone with a license. Because of the risks involved, many of the very hazardous pesticides are labeled "RESTRICTED USE PESTICIDE." Apply such products only by or under the direct supervision of licensed individuals.

Use pesticides safely to protect people and the environment. Begin by properly diagnosing your pest problem. If a pesticide is needed, select the proper one. Always follow all label directions and obey all federal, state, and local laws and regulations concerning pesticides.

Recommendations of specific chemicals are based on information on the manufacturer's label and performance in a limited number of trials. Because environmental conditions and methods of application by growers may vary widely, performance of the chemical will not always conform to the safety and pest control standards indicated by experimental data.

Recommendations for the use of agricultural chemicals are included in this publication as a convenience to the reader. The use of brand names and any mention or listing of commercial products or services does not imply endorsement by North Carolina State University or discrimination against similar products or services not mentioned. Individuals who use agricultural chemicals are responsible for ensuring that the intended use complies with current regulations and conforms to the product label. Be sure to obtain current information about usage regulations and examine a current product label before applying any chemical. N.C. Cooperative Extension agents may also be of assistance.

See the following NC State Extension publications for more information on turfgrass management:

- *Turfgrass Pest Management Manual,* AG-348
  content.ces.ncsu.edu/turfgrass-pest-management-manual
- *North Carolina Agricultural Chemicals Manual*
  content.ces.ncsu.edu/north-carolina-agricultural-chemicals-manual

NC State University Turf Work Group

Faculty Name	Phone Number	E-mail Address	Turf Specialty	Assignment: Teaching (T) Research (R) Extension (E)
Terri **Billeisen**	919.513.2209	tlhoctor@ncsu.edu	Entomology	E
Rick **Brandenburg**	919-515-8876	Rick_Brandenburg@ncsu.edu	Entomology	R/E
Lee **Butler**	919-513-3878	lee_butler@ncsu.edu	Pathology	E
Danesha **Seth-Carley**	919-515-2717	Danesha_Carley@ncsu.edu	Sustainable managed ecosystems	R
Rich **Cooper**	919-515-7600	Rich_Cooper@ncsu.edu	General turf culture	T/R
Emily **Erickson**	919-513-2034	Emily_Erickson@ncsu.edu	Ag Institute	T
Travis **Gannon**	919-513-4655	Travis_Gannon@ncsu.edu	Pesticide fate	R
James **Kerns**	919-513-3878	James_Kerns@ncsu.edu	Pathology	R/E
Matt **Martin**	910-675-2314	Matt_Martin@ncsu.edu	On-site diagnostics	E
Susana **Milla-Lewis**	919-515-3196	Susana_Milla-Lewis@ncsu.edu	Breeding and genetics	T/R
Grady **Miller**	919-515-5656	Grady_Miller@ncsu.edu	Cultivar evaluation/athletic field management	T/R/E
Charles **Peacock**	919-515-7615	Charles_Peacock@ncsu.edu	Web-based applications	T/R/E
Tom **Rufty**	919-515-3660	Tom_Rufty@ncsu.edu	Stress physiology	R
Chadi **Sayde**	919-515-6702	csayde@ncsu.edu	Irrigation	R
Fred **Yelverton**	919-515-5639	Fred_Yelverton@ncsu.edu	Weed management/PGRs	R/E

Commercial Turf Insect Control

R. L. Brandenburg and T. L. Billeisen, Entomology and Plant Pathology

Insect Control in Commercial Turf

Pest	Insecticide and Formulation	Amount per 1,000 sq ft	Precautions and Remarks
Annual Bluegrass Weevil	bifenthrin (Menace, Talstar, others) F, GC	0.25 to 0.5 fl oz	Monitor for adults, apply at peak activity. Use GC formulation for golf courses. Repeated use will lead to resistance issues. Be sure to rotate with other active ingredients to avoid resistance.
	chlorantraniliprole (Acelepryn)	.28 fl oz	Apply approximately 7 to 14 days after adulticide to target larvae.
	cyantraniliprole (Ference)	0.28 fl oz	Monitor for adults, apply at peak activity. Apply approximately 7 to 14 days after adulticide to target larvae.
	indoxacarb (Provaunt) SC	0.28 fl oz	Monitor for adults, apply at peak activity. Apply approximately 7 to 14 days after adulticide to target larvae
	lambda-cyhalothrin (Battle, Scimitar, Cyonara)	0.23 fl oz	Monitor for adults, apply at peak activity.
Ant (also see Imported Fire Ant)	bifenthrin[1] (Menace, Talstar, others) F, GC; G form also available	0.5 to 1 fl oz	Use GC formulation for golf courses.
	carbaryl[1] (Sevin) 80 WSP	1 to 1.5 oz	
	zeta-Cypermethrin, bifenthrin, and imidacloprid (Triple Crown)	20 to 35 fl oz/acre	
	clothianidin + bifenthrin (Aloft) GC SC LC SC GC G LC G	0.27 to 0.54 fl oz 0.27 to 0.54 fl oz 1.8 to 3.6 lb 1.8 to 3.6 lb	
	cyfluthrin (Tempo SC)	0.143 fl oz	Home lawns only.
	cypermethrin[1] (Demon) TC	See label	
	deltamethrin (Deltagard) G	2 to 3 lb/1,000 ft	
	fipronil 0.0143 G (Top Choice, Taurus G)	2 lb	
	hydramethylnon[1] (Maxforce G, Amdro)	See label	
	lambda-cyhalothrin[1] (Battle, Scimitar, Cyonara)	See label	Do not make applications within 20 feet of any body of water. No reentry until spray has dried.
Bee and Wasp (Burrowing)	carbaryl[1] (Sevin) 80 WSP	1.5 oz	
	pyrethroids[1] (Advanced Garden, Battle, Deltagard, Menace, Scimitar, Talstar, Tempo)	See label	
Bermudagrass Mite	abamectin (Divanem)	3.125 to 6.25 fl oz/acre	Tank mix with wetting agent and irrigate 0.1 to 0.25 in water post application. Applicator must be in possession of the 2(ee) label recommendation for restricted uses.
Billbug	bifenthrin[1] (Menace, Talstar, others) F, GC; G form also available	0.25 to 0.5 fl oz	Use GC formulation for golf courses.
	chlorantraniliprole (Acelepryn)	0.184 to 0.46 fl oz	
	chlorpyrifos[1] (Dursban) 50 WSP, Pro	See label	For use on golf courses; check new label.
	clothianidin (Arena) .5G 50 WDG	14 to 22 oz 0.15 to 0.22 oz	
	clothianidin + bifenthrin (Aloft) GC SC LC SC GC G LC G	0.27 to 0.44 fl oz 0.27 to 0.54 fl oz 1.8 to 3.6 lb 1.8 to 3.6 lb	
	deltamethrin (Deltagard) G	2 to 3 lb/1,000 ft	
	dinotefuran (Zylam) 20 SG	1 oz	
	imidacloprid[1] (Merit) 75 WSP	3 to 4 level tsp	Make application prior to egg hatch.
	lambda-cyhalothrin[1] (Battle, Scimitar, Cyonara)	See label	Observe restrictions near water.
	thiamethoxam (Meridian) 0.33 G 25 WG	60 to 80 lb/acre 12.7 to 17 oz/acre	Optimum control when applied from peak flight of adults to peak of egg hatch. Also suppresses mole crickets and chinch bugs.
	zeta-cypermethrin, bifenthrin, and imidacloprid (Triple Crown)	10 to 20 fl oz/acre	
Chinch Bug	acephate[1] (Orthene T, T&O) 75 S	1.2 to 2.4 oz	
	bifenthrin[1] (Menace, Talstar, others) F, GC; G form also available	0.25 to 0.5 fl oz	Use GC formulation for golf courses.
	carbaryl[1] (Sevin) 80 WSP	2.5 to 3 oz	
	chlorantraniliprole (Acelepryn)	0.184 to 0.46 fl oz	Suppression.
	clothianidin (Arena) .5G 50 WDG	1.4 to 1.8 lb 0.2 to 0.3 oz	

Insect Control in Commercial Turf (continued)

Pest	Insecticide and Formulation	Amount per 1,000 sq ft	Precautions and Remarks
Chinch Bug (continued)	clothianidin + bifenthrin (Aloft)		
	GC SC	0.27 to 0.44 fl oz	
	LC SC	0.27 to 0.54 fl oz	
	GC G	1.8 to 3.6 lb	
	LC G	1.8 to 3.6 lb	
	chlorpyrifos[1] (Dursban), 2E, 4E, 50 WP, Pro	See label	For use on golf courses; check new label.
	cyfluthrin (Tempo SC)	0.2 fl oz	Home lawns only.
	cypermethrin (Demon) TC	0.33 to 0.65 fl oz	
	deltamethrin (Deltagard) G	2 to 3 lb/1,000 ft	
	dinotefuran (Zylam) 20 SG	1 oz	For suppression.
	lambda-cyhalothrin[1] (Battle, Scimitar, Cyonara)	See label	Do not make applications within 20 feet of any body of water. No reentry until spray has dried.
	permethrin[1] (Astro)	0.4 to 0.8 fl oz	
	zeta-Cypermethrin, bifenthrin, and imidacloprid (Triple Crown)	20 to 35 fl oz/acre	
Cutworm, Armyworm	acephate[1] (Orthene T, T&O)	1.2 to 2.4 oz	Commercial and residential turf only.
	azadirachtin[1] (Neemix, Turplex)	See label	
	bifenthrin[1] (Menace, Talstar, others) F, GC; G form also available	0.18 to 0.25 fl oz	Use GC formulation for golf courses.
	Bt products, various labels	See label	
	carbaryl[1] (Sevin) 80 WSP and baits	0.75 to 1.5 oz	Treat in late afternoon. Apply in adequate water for good coverage but do not flood or water in. Do not cut grass for 1 to 3 days after treatment.
	chlorantraniliprole (Acelepryn)	0.046 to 0.092 fl oz	
	chlorpyrifos[1] (Dursban) 4 E, 2 ES, 50 WP, Pro	See label	For use on golf courses; check new label.
	clothianidin (Arena)		Cutworms only.
	.5G	1.4 to 1.8 lb	
	50 WDG	0.2 to 0.3 oz	
	clothianidin + bifenthrin (Aloft)		
	GC SC	0.27 to 0.54 fl oz	
	LC SC	0.27 to 0.54 fl oz	
	GC G	1.8 to 3.6 lb	
	LC G	1.8 to 3.6 lb	
	cyfluthrin[1] (Tempo SC)	0.143 fl oz	Home lawns only.
	deltamethrin (Deltagard) G	2 to 3 lb/1,000 ft	
	dinotefuran (Zylam) 20 SG	1 oz	
	entomogenous nematodes[1]	See label	Read and follow special application instructions. Effective only against small cutworms.
	indoxacarb (Provaunt) SC	0.0625 to 0.25 fl oz	Not labeled for use on sod farms.
	lambda-cyhalothrin[1] (Battle, Scimitar, Cyonara)	See label	Do not make applications within 20 feet of any body of water. No reentry until spray has dried.
	spinosad A + D (Conserve) SC	1.25 fl oz	Rate varies with size and species.
	tetraniliprole (Tetrino)	0.367 to 0.735 fl oz	Apply when pest presence first observed or anticipated
	trichlorfon (Dylox, Proxol) 80 SP	1.5 to 3 oz	
Earthworm			Usually not a problem. No effective controls available.
Fall Armyworm	acephate[1] (Orthene, T, T&O)	0.5 to 1.2 oz	Water in immediately after application.
	chlorantraniliprole (Acelepryn)	0.046 to 0.092 fl oz	
	chlorpyrifos[1] (Dursban) 4 E, 2 E, 50WP, Pro	See label	For use on golf courses; check new label.
	indoxacarb (Provaunt) SC	0.0625 to 0.25 fl oz	Not labeled for use on sod farms.
	pyrethroids[1] (Advanced Garden, Battle, Deltagard, Menace, Scimitar, Talstar, Tempo, Cyonara)	See label	
	spinosad A + D (Conserve SC)	1.25 fl oz	Rate varies with size and species.
	tetraniliprole (Tetrino)	0.367 to 0.735 fl oz	Apply when pest presence first observed or anticipated
Grasshopper	acephate[1] (Orthene T, T&O)	0.5 oz	Do not mow turfgrass for at least 24 hours after application.
	deltamethrin (Deltagard) G	2 to 3 lb/1,000 ft	
	lambda-cyhalothrin[1] (Battle, Scimitar, Cyonara)	See label	Do not make applications within 20 feet of any body of water. No reentry until spray has dried.
Ground Pearl			No effective control—practice good management.

Insect Control in Commercial Turf (continued)

Pest	Insecticide and Formulation	Amount per 1,000 sq ft	Precautions and Remarks
Imported Fire Ant (See www.ncagr.gov/ plantindustry/plant/ entomology/documents/ ncifaquarantine.pdf for latest quarantine areas.)	acephate[1] (Lesco-Fate) (Orthene, T, T&O) 75 S	See label 1 to 2 tsp/mound	Distribute uniformly over mound. For best results apply in early morning or late afternoon.
	bifenthrin[1] (Menace, Talstar, others) F; G form also available		Follow label directions.
	clothianidin + bifenthrin (Aloft) GC SC LC SC GC G LC G	See label 0.27 to 0.44 fl oz 0.27 to 0.54 fl oz 1.8 to 3.6 lb 1.8 to 3.6 lb	
Imported Fire Ant (See www.ncagr.gov/ plantindustry/plant/ entomology/documents/ ncifaquarantine.pdf for latest quarantine areas.)	deltamethrin (Deltagard) G	2 to 3 lb/	
	fenoxycarb (Award)[1] B	1 to 3 level tbsp 1 to 1.5 lb/acre	Single mound treatment. Apply uniformly with ground equipment.
	fipronil (Topchoice, Fipronil, others) 0.0143	2 lb	Apply as a broadcast.
	fipronil + bifenthrin + lambda-cyhalothrin (Taurus Trio G)	2 lb	Apply as a broadcast. Irrigate prior to treatment.
	hydramethylnon[1] (Amdro) 0.88% bait (Maxforce G)	— See label	Uniformly broadcast 1 to 1.5 pound of bait per acre with ground equipment on pastures, range grasses, lawns, and nonagricultural lands. Or distribute uniformly 5 level tablespoons of bait 3 to 4 feet around base of each mound. Do not exceed 1.5 pounds per acre.
	imidacloprid + bifenthrin (Allectus, Atera)	See label	Rate varies with pest. Different formulations for different sites.
	indoxacarb (Advion) bait	1.5 lb/acre	Bait formulation.
	lambda-cyhalothrin[1] (Battle, Scimitar, Cyonara)	See label	
	metaflumizone (Siesta) bait	1.0 to 1.5 lb/acre 2 to 4 tbsp/mound	Do not exceed 4 applications in a one-year period.
	methoprene (Extinguish) 0.5 % bait	1.5 lb/acre	Mound or broadcast.
	methoprene + hydramethylnon (Extinguish Plus)	1.5 lb/acre	
	pyriproxyfen (Distance, Esteem)	See label	Mound or broadcast.
	spinosad (Justice bait)	See label	
	spinosad A + D (Conserve SC)	0.1 fl oz/gal/mound	Dilute 0.1 fluid ounce in 1 gallon water. Use 1 to 2 gallons per mound.
Leafhopper, Spittlebug	acephate[1] (Orthene, T, T&O) 75 S	1 oz	
	bifenthrin[1] (Menace, Talstar, others) F, GC; G form also available	0.25 to 0.5 fl oz	Use GC formulation for golf courses.
	carbaryl[1] (Sevin) 80 WSP	0.75 to 1.5 oz	
	chlorpyrifos[1] (Dursban) 4 E, 50 WSP, Pro	See label	For use on golf courses; check new label.
	deltamethrin (Deltagard) G	2 to 3 lb	
Millipede	bifenthrin[1] (Menace, Talstar, others) F, GC; G form also available	0.25 to 0.5 fl oz	Use GC formulation for golf courses.
	carbaryl[1] (Sevimol) (Sevin) 80 WSP	1.5 to 3 oz 0.75 to 1.5 oz	
	chlorpyrifos[1] (Dursban) 2 E, Pro	See label	For use on golf courses; check new label.
	cypermethrin (Demon) TC	See label	
	lambda-cyhalothrin[1] (Battle, Scimitar, Cyonara)	See label	Do not make applications within 20 ft of any body of water. No reentry until spray has dried.
Mole Cricket	acephate[1] (Orthene T, T&O, Lesco-Fate)	1 to 1.9 oz	Water soil before application. Do not water in.
	bifenthrin[1] (Menace, Talstar, others) F, GC; G form also available	0.5 to 1 fl oz	Use GC formulation for golf course.
	carbaryl[1] (Sevin) baits	See label	
	cyfluthrin[1] (Tempo SC, Tempo Ultra)	0.2 fl oz	Home lawn use only.
	deltamethrin (Deltagard) G	2 to 3 lb	
	dinotefuran (Zylam) 20 SG	See label	
	entomogenous nematodes[1]	See label	Various formulations now available. Adequate soil moisture critical for good control.
	fipronil (Chipco Choice, others) 0.1 G (Top Choice, Fipronil, others) 0.0143	12.5-25 lb/A 2 lb	Use slit placement equipment. Apply as a broadcast.
	imidacloprid (Merit) 75 WP 0.5G	4 level tsp 1.8 lb	Apply while crickets are less than ½ inch long (June, early July).
	indoxacarb (Advion) Insect G	50 to 200 lb/acre	Not for use on sod farms. DO NOT water in after application.
	indoxacarb (Provaunt)	0.275 oz	Two applications 2 to 4 weeks apart work best, following egg hatch.
	lambda-cyhalothrin[1] (Battle, Scimitar, Cyonara)	See label	Do not make applications within 20 feet of any body of water. No reentry until spray has dried.
	zeta-Cypermethrin, bifenthrin, and imidacloprid	20 to 35 fl oz/acre	
Slug, Snail	mesurol 2 B	1 lb	Apply late in afternoon.
	metaldehyde	See label	

Insect Control in Commercial Turf (continued)

Pest	Insecticide and Formulation	Amount per 1,000 sq ft	Precautions and Remarks
Sod Webworm	acephate[1]		
	(Lesco-Fate, Orthene T, T&O)	0.5 to 1 oz	Home lawns only.
	(Precise 4G)	2.8 lb	Irrigate immediately.
	azadirachtin[1] (Azatrol, Neemix, Turplex)	0.5 fl oz	
	Bacillus thuringiensis, various brands	1 to 2 lb/acre	
	bifenthrin[1] (Menace, Talstar, others) F, GC; G form also available	0.18 to 0.25 fl oz	Use GC formulation for golf courses.
	carbaryl[1] (Sevin) 80 WSP	2.5 to 3 oz	
	chlorantraniliprole (Acelepryn)	0.046 to 0.092 fl oz	
	chlorpyrifos[1] (Dursban) 4 E, 2 E, 5 G, Pro	See label	For use on golf courses; check new label.
	clothianidin (Arena)		
	.5G	14 to 22 oz	
	50 WDG	0.15 to 0.22 oz	
Sod Webworm (continued)	clothianidin + bifenthrin (Aloft)		
	GC SC	0.27 to 0.54 fl oz	
	LC SC	0.27 to 0.54 fl oz	
	GC G	1.8 to 3.6 lb	
	LC G	1.8 to 3.6 lb	
	cyfluthrin[1] (Tempo SC, Tempo Ultra)	0.143 fl oz	Irrigate immediately after application. Do not apply to newly seeded stands or bentgrass.
	deltamethrin (Deltagard) G	2 to 3 lb	
	dinotefuran (Zylam) 20 SG	1 oz	
	indoxacarb (Provaunt) SC	0.0625 to 0.25 fl oz	Not labeled for use on sod farms.
	lambda-cyhalothrin[1] (Cyonara, Scimitar, Battle)	See label	Do not make applications within 20 feet of any body of water. No reentry until spray has dried.
	permethrin[1] (Astro)	0.4 to 0.8 fl oz	
	spinosad A + D (Conserve) SC	1.25 fl oz	Rate varies with size and species.
	tetraniliprole (Tetrino)	0.367 to 0.735 fl oz	Apply when pest presence first observed or anticipated
	trichlorfon[1] (Dylox, Proxol) 80 SP	1.5 to 3 oz	
Sowbug, Pillbug	bifenthrin[1] (Talstar) F, GC G form also available	0.25 to 0.5 fl oz	Use GC formulation for golf courses.
	carbaryl[1] (Sevin) 80 WSP	0.75 to 1.5 oz	
	cypermethrin[1] (Demon) TC	See label	
	deltamethrin (Deltagard) G	2 to 3 lb	
	lambda-cyhalothrin[1] (Battle, Cyonara, Scimitar)	See label	Do not make applications within 20 feet of any body of water. No reentry until spray has dried.
Sugarcane Beetle	bifenthrin[1] (Talstar) F, GC G form also available	0.5 to 1.0 fl oz	Target adults early (Apr-May). Insecticide efficacy significantly reduced for fall population.
White Grub (May beetle, chafers, green June beetle, and others)	*B.t.* subspecies galleriae (grubGoneG)	100 to 150 lbs/acre	
	chlorantraniliprole (Acelepryn)	0.184 to 0.367 fl oz	Optimal control when applied at egg hatch. Use higher rates later in summer.
	clothianidin (Arena)		Mole cricket suppression.
	.5G	14 to 22 oz	
	50 WDG	0.15 to 0.22 oz	
	clothianidin + bifenthrin (Aloft)		
	GC SC	0.27 to 0.54 fl oz	
	LC SC	0.27 to 0.54 fl oz	
	GC G	1.8 to 3.6 lb	
	LC G	1.8 to 3.6 lb	
	dinotefuran (Zylam) 20 SG	1 oz	
	imidacloprid[1] (Merit) 75 WP	3 to 4 level tsp	Make application prior to egg hatch. (Offers some suppression of caterpillars.)
	thiamethoxam (Meridian)		Optimum control when applied from peak flight of adults to peak of egg hatch. Also suppresses mole crickets and chinch bugs.
	0.33 G	60 to 80 lb/acre	
	25 WG	12.7 to 17 oz/acre	
	trichlorfon (Dylox, Proxol) 80 SP	3.75 oz	Can be used with some success as a rescue treatment in August and September. Apply at egg hatch.

Insect Control in Commercial Turf (continued)

Pest	Insecticide and Formulation	Amount per 1,000 sq ft	Precautions and Remarks
White Grub, Green June Beetle (only)	*B.t.* subspecies galleriae (grubGoneG)	100 to 150 lb/acre	
	carbaryl[1] (Sevin) 80 WSP	1 to 1.5 oz	
	chlorantraniliprole (Acelepryn)	0.184 to 0.367 fl oz	Optimal control when applied at egg hatch. Use higher rates later in summer.
	chlorpyrifos[1] (Dursban) 50 WSP, Pro	See label	For use on golf courses; see new label.
	clothianidin (Arena)		Mole cricket suppression.
	.5G	14 to 22 oz	
	50 WDG	0.15 to 0.22 oz	
	clothianidin + bifenthrin (Aloft)		
	GC SC	0.27 to 0.54 fl oz	
	LC SC	0.27 to 0.54 fl oz	
	GC G	1.8 to 3.6 lb	
	LC G	1.8 to 3.6 lb	
	dinotefuran (Zylam) 20 SG	1 oz	Apply at egg hatch.
	imidacloprid[1] (Merit) 75 WP	3 to 4 level tsp	Make application prior to egg hatch. Do not use on sod farms. Offers some suppression of caterpillars.
	thiamethoxam (Meridian)		Optimum control when applied from peak flight of adults to peak of egg hatch. Also suppresses mole crickets and chinch bugs.
	0.33 G	60 to 80 lb/acre	
	25 WG	12.7 to 17 oz/acre	
White Grub (Japanese beetle)	*B.t.* subspecies galleriae (grubGoneG)	100 to 150 lb/acre	
	carbaryl[1] (Sevin) 80 WSP	3 oz	
	chlorantraniliprole (Acelepryn)	0.184 to 0.367 fl oz	Optimal control when applied at egg hatch. Use higher rates later in summer.
	clothianidin + bifenthrin (Aloft)		
	GC SC	0.27 to 0.54 fl oz	
	LC SC	0.27 to 0.54 fl oz	
	GC G	1.8 to 3.6 lb	
	LC G	1.8 to 3.6 lb	
	clothianidin (Arena)		Mole cricket suppression.
	.5G	14 to 22 oz	
	50 WDG	0.15 to 0.22 oz	
	dinotefuran (Zylam) 20SG	1 oz per 1000 sq ft	Can be used with some success as a rescue treatment in August and September. Apply at egg hatch.
	imidacloprid[1] (Merit) 75 WP	3 to 4 level tsp	Make application prior to egg hatch. (Offers some suppression of caterpillars.)
	zeta-Cypermethrin, bifenthrin, and imidacloprid (Triple Crown)	20 to 35 fl oz/acre	
	thiamethoxam (Meridian)		Optimum control when applied from peak flight of adults to peak of egg hatch. Also suppresses mole crickets and chinch bugs.
	0.33 G	60 to 80 lb/acre	
	25 WG	12.7 to 17 oz/acre	
	trichlorfon[1] (Dylox, Proxol) 80 SP	3.75 oz	Can be used with some success as a rescue treatment in August and September. Apply at egg hatch.
Zoysiagrass mites			No effective controls available. Make sure turf is irrigated and apply lime and fertilizer applications according to soil test recommendations.

[1]Several tradenames available. Check label for active ingredient. Always follow label instructions.

Chemical Weed Control in Lawns and Turf

F. H. Yelverton, T. W. Gannon, and G.L. Miller, Crop and Soil Sciences Department

Note: A mode of action code has been added to the Herbicide and Formulation column of this table. Use MOA codes for herbicide resistance management. See Herbicide Modes of Action for Hay Crops, Lawns and Turf for details concerning active ingredients, brand names, chemical families and modes of action.

Several of the preemergence herbicides are available on fertilizer carriers for homeowner application.

Chemical Weed Control in Lawns and Turf

Herbicide and Formulation	Brand Names	Amount of Formulation per 1,000 sq ft	Amount of Formulation per Acre	Pounds Active Ingredient per Acre	Precautions and Remarks
Preemergence Control, Smooth and Large Crabgrass, Goosegrass, Foxtails, other annual grasses					
benefin, MOA 3 (2.5 G)	Lebanon Balan 2.5 G, The Andersons Crabgrass Preventer with 2.5 Balan	2.75 lb	120 lb	3	Safe to apply to all established turfgrass except bentgrass. Do not apply in the spring to lawns seeded the previous fall or to golf course greens. Do not use on newly sprigged turfgrasses.
[benefin + trifluralin], MOA 3 + 3 (0.86 G)	Fertilizer with Team Pro 0.86%	8 lb	349 lb	3	Use on lawns and golf course fairways of bahiagrass, bentgrass, bermudagrass, centipedegrass, fescue, perennial ryegrass, St. Augustinegrass, and zoysiagrass.
bensulide, MOA 8 (4 EC) (8.5 G, 12.5 G)	(4 EC): Bensumec 4 LF (8.5 G): Weedgrass Preventer (12.5 G): Pre-San Granular		Varies, several concentrations available	10	May be applied to all established turfgrass and dichondra, residential lawns, and golf course greens and tees. Limit 2 applications per year to greens and tees. Do not use on newly sprigged turfgrasses. Not effective for goosegrass control.
[bensulide + oxadiazon], MOA 8 + 14 (6.56 G)	Goosegrass / Crabgrass Control	2.6 lb	116 lb	6 + 1.5	Controls crabgrass and goosegrass. Use on established bermudagrass, zoysiagrass, tall fescue, bentgrass, perennial bluegrass, or perennial ryegrass fairways and tees. Use also on bermudagrass and bentgrass greens.
dithiopyr, MOA 4 (2 EW, 2 L) (40 WP)	(2 EW, 2 L): Armortech CGC 2 L, Dimension 2 EW, Dithiopyr 2 L (40 WP): Armortech CGC 40, Dimension Ultra, Dithiopyr 40 WSP	0.75 fl oz 0.46 oz	1 qt 20 oz	0.5	May be applied to most all cool-season and warm-season turfgrasses except colonial bentgrass. See label for injury precautions regarding certain varieties. Also controls pre-tillered crabgrass. Split applications recommended in southern and coastal regions of the state (0.25 pound a.i. at 8-week intervals). Timely irrigation or rainfall is critical for activation.
indaziflam, MOA 21 (20 WSP)	(20 WSP): Specticle 20 WSP	0.057 to 0.115 oz	2.5 to 5 oz	0.03125 to 0.0625	Use only on established turf (1 year after seeding) such as bermudagrass, zoysiagrass, centipedegrass, St. Augustinegrass, seashore paspalum, and bahiagrass. Labeled for commercial and residential lawns, golf courses (roughs, tees, fairways), sod farms, athletic fields, parks and cemeteries. Use a minimum of 2.5 ounces per acre for crabgrass, annual bluegrass and broadleaf weed control and a minimum of 3.75 ounces per acre for goosegrass, annual sedge and kyllinga species control. Apply up to 2.5 ounces per acre on centipedegrass and St. Augustinegrass due to tolerance concerns. For all other tolerant turfgrasses, do not exceed 5 ounces per acre in a single application or 7.1 ounces per acre within a calendar year. There is an 8 month overseeding restriction following a 2.5 ounces per acre application. Can sprig 2 months following application, or if sprigged first, wait 4 months before spraying. Can sod 4 months following application, or if sodded first wait 2 months after rooting before spraying.
(0.622F)	(0.622F): Specticle Flo	0.69 to 0.23 fl oz	3 to 10 fl oz	0.01458 to 0.0486	Use up to 6 fluid ounces per acre on common bermudagrass, centipedegrass and St. Augustinegrass and 10 fluid ounces per acre on hybrid bermudagrass, zoysiagrass and bahiagrass established 16 months in areas such as golf course roughs and fairways, residential and commercial turf, sod farms, athletic fields, parks and cemeteries. 10 fluid ounces per acre needed for annual sedge and kyllinga species control. Don't exceed 18.5 fluid ounces per acre per year. Do not vertical mow 1 month before or after application. Irrigate within 2 days of treatment for maximum benefit. Check label for split or multiple application rates and timings. Delay overseeding 10 months if 4.5 to 6 fluid ounces used and 12 months if 6 to 9 fluid ounces used. For sod production, only apply to bermudagrass, zoysiagrass or bahiagrass. Apply if 80% ribbon coverage and before 4 months prior to harvest. Wait 6 month after treatment if sodding bare ground. Apply to actively growing sod established for 3 months.
(0.0224 G)	(0.0224 G): Specticle G	2.9 to 4.6 lb	125 to 200 lb	0.028 to 0.045	Use on same warm season turf species established at least 16 months and sites as above. Do not exceed 400 pounds product per year. Allow a 15 feet buffer from cool season turf areas. Do not apply upslope from cool season turf.
metolachlor, MOA 15 (7.62 EC)	Pennant Magnum	0.96 fl oz	2.6 pt	2.48	Apply to established bermudagrass, centipedegrass, St. Augustinegrass, bahiagrass, and zoysiagrass. Can apply up to 4.2 pints per acre per year to same area used for commercial sod production.

Chemical Weed Control in Lawns and Turf (continued)

Herbicide and Formulation	Brand Names	Amount of Formulation per 1,000 sq ft	Amount of Formulation per Acre	Pounds Active Ingredient per Acre	Precautions and Remarks
Preemergence Control, Smooth and Large Crabgrass, Goosegrass, Foxtails, other annual grasses (continued)					
napropamide, MOA 15 (50 DF)	Devrinol 50 DF	1.5 to 2.2 oz	4 to 6 lb	2 to 3	Use in established bahiagrass, bermudagrass, centipedegrass, St. Augustinegrass, and tall fescue.
oryzalin, MOA 3 (4 AS, 4 L)	(4 AS, 4 L): Monterey Weed Impede 4 AS, Oryzalin 4 AS, Phoenix Harrier 4 L, Surflan AS	1.5 fl oz	2 qt	2	Use on established bahiagrass, centipedegrass, tall fescue, St. Augustinegrass, zoysiagrass, and bermudagrass except greens and tees. A total of 3 quart per acre may be used if application is split by applying 1.5 quarts per acre followed by 1.5 quarts per acre 8 to 10 weeks later. Follow label directions. Do not apply in the spring or summer to tall fescue reseeded the previous fall.
oryzalin, MOA 3 (85 WDG)	(85WDG): Surflan WDG	0.64 to 0.88 oz	1.75 to 2.4 lb	1.4875 to 2.04	Observe same turf tolerances and tall fescue precautions as above. Successful preemergence activity should occur if activated by 0.5 inch of water within 21 days of application. Apply 2.4 pounds per acre as a single application or 1.75 pounds per acre in sequential applications spaced 12 weeks apart.
oxadiazon, MOA 14 (2 G)	(2 G): Oxadiazon 2 G, Ronstar G	2.3 to 4.6 lb	100 to 200 lb	2 to 4	Use in established perennial bluegrass, perennial ryegrass, bentgrass, bermudagrass, tall fescue, zoysiagrass, and St. Augustinegrass. Red fescue is not tolerant. Do not apply to dichondra, centipedegrass, putting greens or tees, or to newly seeded areas. Do not apply to bentgrass mowed at less than 3/8 inch. Do not apply to wet turf. Rainfall or irrigation after application will improve weed control activity. May be applied when sprigging bermudagrass and zoysiagrass. Do not apply to home lawns.
oxadiazon, MOA 14 (50 WP)	(50 WP): Oxadiazon 50 WSB, Ronstar 50 WSB	1.5 to 2.2 oz	4 to 6 lb	2 to 3	Use in dormant, established bermudagrass, St. Augustinegrass, and zoysiagrass in fairways and parks. Should be applied at least 2 to 3 weeks before greenup of turf. May be applied when sprigging bermudagrass and zoysiagrass. Do not use on home lawns.
oxadiazon, MOA 14 (3.17 SC)	(3.17 SC): Oxadiazon SC, Phoenix Starfighter L, Ronstar Flo	1.85 to 2.8 fl oz	2.52 to 3.81 qt	2 to 3	Use in dormant, established bermudagrass, St. Augustinegrass, and zoysiagrass in fairways and parks. May apply 2 lb a.i. per acre when sprigging bermudagrass. Apply at least 2 to 3 weeks before greenup of turf. Do not use on home lawns.
[oxadiazon + prodiamine], MOA 14 + 3 (1.2 G)	Pro-mate Ronstar + Barricade 1.2 G, Regalstar II	4.5 lb	200 lb	2 + 0.4	Use on turf, golf courses (excluding putting greens) of established bermudagrass, zoysiagrass, St. Augustinegrass, ryegrass, centipedegrass, bentgrass, bluegrass, and tall fescue. Contains 38% N. Apply to dry foliage.
pendimethalin, MOA 3 (2 G)	(2 G): Pendulum 2 G	1.72 to 3.44 lb	75 to 150 lb	1,5 to 3	Use on established bahiagrass, bermudagrass, centipedegrass, fine fescue, Kentucky bluegrass, perennial ryegrass, St. Augustinegrass, tall fescue, and zoysiagrass. Do not use on winter-overseeded grasses. Wait 4 months after treatment to seed or sod. Do not apply to newly seeded turf until after the fourth mowing. Do not apply to newly sprigged turf until 5 months establishment.
0.86 G	(0.86 G): fertilizers – Pendimethalin, Pre-M, Propendi, Pro-mate Pendi	2.67 to 5.34 lb	116 to 232 lb	1 to 2	
(1.29 G)	(1.29 G): Step 1 Crabgrass Preventer, Turf Builder with Halts	2.67 lb	116 lb	1.5	
pendimethalin, MOA 3 (3.8 CS)	(3.8 CS): HydroCap, Pendulum AquaCap, Pre-M AquaCap, Prowl H2O, Satellite HydroCap	1.15 to 2.3 fl oz	3.1 to 6.3 pt	1.5 to 3	Use on noncropland as well as established nonresidential and residential turf areas mowed at least 4 times consisting of bahiagrass, bermudagrass, buffalograss, centipedegrass, St. Augustinegrass, zoysiagrass, Kentucky bluegrass, perennial ryegrass, bentgrass, established *Poa annua* (0.5 inch height or taller), fine fescue, and tall fescue. Do not use on bentgrass or *Poa annua* greens and tees. If lower rate is applied initially, repeat in 6-8 weeks for extended control. Do not reseed or overseed into treated turfgrass for 3 months, or sprig turfgrass for 5 months following application. Do not exceed 4.2 pints per acre on residential and sod farm turfgrass.
[pendimethalin + dimethemamid], MOA 3 + 15 (1.75 G)		2.3 to 4.6 lb	100 to 200	1.75 to 3.5	Use on residential, commercial, recreational, sod farm and golf course turf, excluding greens. Tolerant turf species include bermudagrass, centipedegrass, St. Augustinegrass, seashore paspalum and zoysiagrass. For extended control, make sequential applications within 5 to 8 weeks not to exceed 400 pounds per acre. Irrigate within 24 hours of application for optimum control. Following application, wait 3 months to overseed, reseed or sprig. If sprigged first, wait 2 months for root establishment to treat. On new sod, mow at least twice before application. On new seedlings, mow at least 4 times before application. Wait 2 weeks after aerification or verticutting before applying.

Chemical Weed Control in Lawns and Turf (continued)

Herbicide and Formulation	Brand Names	Amount of Formulation per 1,000 sq ft	Amount of Formulation per Acre	Pounds Active Ingredient per Acre	Precautions and Remarks
Preemergence Control, Smooth and Large Crabgrass, Goosegrass, Foxtails, other annual grasses (continued)					
prodiamine, MOA 3 (65 WG)	(65 WDG): Armortech Kade, Barricade, Calvacade, Halts Pro, Phoenix Knighthawk, ProClipse, Prodiamine, Quali-Pro Prodiamine, RegalKade, Resolute, Stonewall	0.185 to 0.83 oz	0.5 to 2.3 lb	0.325 to 1.5	May be used on established bahiagrass, bermudagrass, centipedegrass, St. Augustinegrass, zoysia, tall fescue, creeping red fescue, perennial bluegrass and ryegrass, and creeping bentgrass. Do not apply to greens. May apply when sprigging or plugging bermudagrass, up to 0.8 pound product per acre.
(4 FL)	(4 FL): Barricade, Evade, Resolute	0.23 to 1.1 fl oz	0.625 to 3 pt	0.3125 to 1.5	
prodiamine, MOA 3 (0.5 G)	(0.5 G): RegalKade, Signature Crabgrass Preventer, Turf Pride	1.5 to 6.9 lb	64 to 300 lb	0.32 to 1.5	See precautions for prodiamine 65 WG and 4 FL above except may be used on established turf only. Do not apply more than 150 pounds per acre per application. Do not make more than two applications per calendar year. Wait at least 60 days after initial application before making a second application. Prodiamine is coated on a 32-3-12 dry fertilizer carrier.
siduron, MOA 7 (50 WP)	Tupersan	7.3 oz	20 lb	10	Use only on bluegrass, fescue, perennial ryegrass, and certain bentgrasses (check label). Can be used at the rate of 8 pounds of formulation when seeding bentgrass, bluegrass, fescue, and ryegrass. Can also be used in newly sprigged or established zoysia. Do not use on bermudagrass, carpetgrass, or centipedegrass.
Preemergence Control, Goosegrass					
dimethenamid, MOA 15 (6 L)	Tower	0.48 to 0.73 fl oz	21 to 32 fl oz	1 to 1.5	Use on residential, commercial, recreational, sod farm and golf course turf, excluding greens. Apply 21 ounces to established bentgrass, bluegrass species, fescue species and perennial ryegrass maintained at 0.5 inch cut, but expect yellowing and stand reduction. Apply 32 ounces to bahiagrass, bermudagrass species, centipedegrass, St. Augustinegrass, seashore paspalum and zoysiagrass. For extended control, make sequential applications within 5 to 8 weeks at 32 fluid ounces per acre rate. Irrigate within 24 hours of application for optimum control. Following application, wait 6 weeks to overseed or reseed, wait 2 months to sprig, wait until 2 mowings for new sod, and wait until 4 mowings for newly seeded turf.
Preemergence Control, Annual Bluegrass (*Poa annua*)					
[benefin + trifluralin], MOA 3 + 3 (0.86 G)	Fertilizer with Team Pro 0.86%	4 to 8 lb	174 to 349 lb	1.5 to 3	Apply during late summer before *Poa annua* germinates. Do not apply to turf areas that are to be overseeded.
bensulide, MOA 8 (4 EC)	(4 EC): Bensumec 4 LF		several concentrations available	12.5	See section on preemergence control of crabgrass and goosegrass or product labels for turfgrass tolerance, precautions and remarks for the listed preemergence annual bluegrass herbicides.
(8.5 G, 12.5 G)	(8.5 G): Weedgrass Preventer				
	(12.5 G): Pre-San Granular				
dithiopyr, MOA 4 (2 EW, 2 L)	(2 EW, 2 L): Armortech CGC 2 L, Dimension 2 EW, Dithiopyr 2 L	0.75 fl oz	1 qt	0.5	Timely irrigation or rainfall is critical for activation.
(40 WP)	(40 WP): Armortech CGC 40, Dimension Ultra, Dithiopyr 40 WSP	0.46 oz	20 oz		
indaziflam, MOA 21 (20 WSP)	(20 WSP): Specticle 20 WSP	0.057 to 0.115 oz	2.5 to 5 oz	0.031 to 0.063	
(0.622 F)	(0.622F): Specticle Flo	0.138 to 0.23 fl oz	6 to 10 fl oz	0.029 to 0.049	
(0.0224 G)	(0.0224 G): Specticle G	2.9 to 4.6 lb	125 to 200 lb	0.028 to 0.045	
metolachlor, MOA 15 (7.62 EC)	Pennant Magnum	0.48 to 0.96 fl oz	1.3 to 2.6 pt	1.24 to 2.48	
napropamide, MOA 15 (50 DF)	Devrinol 50 DF	1.5 to 2.25 oz	4 to 6 lb	2 to 3	
oryzalin, MOA 3 (4 AS)	Oryzalin 4 AS, Phoenix Harrier 4 L, Surflan AS	1.1 fl oz	1.5 qt	1.5	Apply full rate unless potentially thin turfgrass cover is a problem caused by dense poa infestation.
(85 WDG)	(85WDG): Surflan WDG	0.64 to 0.88 oz	1.75 to 2.4 lb	1.4875 to 2.04	
oxadiazon, MOA 14 (2 G)	(2 G): Oxadiazon 2 G, Ronstar G	2.3 to 4.6 lb	100 to 200 lb	2 to 4	

Chemical Weed Control in Lawns and Turf (continued)

Herbicide and Formulation	Brand Names	Amount of Formulation per 1,000 sq ft	Amount of Formulation per Acre	Pounds Active Ingredient per Acre	Precautions and Remarks
Preemergence Control, Annual Bluegrass (*Poa annua*) (continued)					
pendimethalin, MOA 3 (2 G)	(2 G): Pendulum 2 G	1.72 to 3.44 lb	75 to 150 lb	1.5 to 3	
(0.86 G)	(0.86 G): fertilizers – Pendimethalin, Pre-M, Propendi, Pro-mate Pendi	2.67 to 5.34 lb	116 to 232 lb	1 to 2	
(1.29 G)	(1.29 G): Step 1 Crabgrass Preventer, Turf Builder with Halts	2.67 lb	116 lb	1.5	
(3.8 CS)	(3.8 CS): HydroCap, Pendulum AquaCap, Pre-M AquaCap, Prowl H2O, Satellite HydroCap	1.15 to 1.55 fl oz	3.1 to 4.2 pt	1.5 to 2	
[pendimethalin + dimethemamid], MOA 3 + 15 (1.75 G)		2.3 to 4.6 lb	100 to 200	1.75 to 3.5	
prodiamine, MOA 3 (65 WG or WDG)	(65 WDG): Armortech Kade, Barricade, Calvacade, Halts Pro, Phoenix Knighthawk, ProClipse, Prodiamine, Quali-Pro Prodiamine, RegalKade, Resolute, Stonewall	0.185 to 0.83 oz	0.5 to 2.3 lb	0.325 to 1.5	
(4 FL)	(4 FL): Barricade, Evade, Resolute	0.23 to 1.1 fl oz	0.625 to 3 pt	0.3125 to 1.5	
pronamide, MOA 3 (3.3 SC)	(3.3 SC): Kerb SC T&O, Willowood Pronamide 3.3SC	0.46 to 1.29 fl oz	1.25 to 3.5 pt	0.5 to 1.5	Not for home use. Can be applied from Sept. 15 to Feb. 1 for preemergence or postemergence annual bluegrass control in bermudagrass, zoysiagrass, centipedegrass, and St. Augustinegrass grown for sod, nonresidential or industrial sites, golf course turf, and stadium or professional athletic fields. 1.25 to 2.5 pints per acre provides preemergence to pre tiller stage control. 2 to 2.5 pints per acre provides postemergence control from early tiller to early seedhead stage. 2.5 to 3.5 pints per acre needed for postemergence control at seedhead stage. Henbit and chickweed species controlled at preemergence timings. Can be used for removal of overseeded grasses; therefore, do not overseed if it is desired to maintain a stand. Do not overseed treated area within 90 days of treatment. Injury symptoms from postemergence applications can take up to 5 weeks to develop.
Preemergence Control, Annual Bluegrass in Overseeded Bermudagrass					
benefin, MOA 3 (2.5 G)	Lebanon Balan 2.5 G, The Andersons Crabgrass Preventer with 2.5 Balan	2.75 lb	120 lb	3	Apply in late summer before *Poa annua* germinates. Perennial ryegrass can be overseeded 6 weeks after benefin is applied.
dithiopyr, MOA 4 (2 EW, 2 L)	(2 EW, 2 L): Armortech CGC 2 L, Dimension 2 EW, Dithiopyr 2 L	0.75 fl oz	1 qt	0.5	Apply in late summer before *Poa annua* germinates. Perennial ryegrass can be overseeded 6 to 8 weeks after application. Apply only on well-established bermudagrass. Do not reapply in fall or winter after overseeding unless injury can be tolerated.
(40 WP)	(40 WP): Armortech CGC 40, Dimension Ultra, Dithiopyr 40 WSP	0.46 oz	20 oz		
prodiamine, MOA 3 (65 WG or 65 WDG)	(65 WDG): Armortech Kade, Barricade, Calvacade, Halts Pro, Phoenix Knighthawk, ProClipse, Prodiamine, Quali-Pro Prodiamine, RegalKade, Resolute, Stonewall	0.213 to 0.367 oz	0.58 to 1 lb	0.37 to 0.65	Use on golf courses (excluding putting greens) when overseeding with perennial ryegrass at a minimum seeding rate of 350 pounds per acre. Apply 8 to 10 weeks before overseeding and expect 70% or greater control. For best potential control, use higher rate and shorter time interval before overseeding. However, this could increase ryegrass seedling mortality or temporarily reduce root growth.
Preemergence and Postemergence Control, Annual Bluegrass					
ethofumesate, MOA 8 (1.5 EC)	(1.5 EC): Progress	2 fl oz	2.67 qt	1	For control of annual bluegrass in dormant bermudagrass overseeded with perennial ryegrass or in established perennial ryegrass turf. Rates are per application. The first application should be 30 to 45 days after overseeding with perennial ryegrass. The second application should be 21 to 28 days later. Do not apply ethofumesate to overseeded bermudagrass after Jan. 1 in N.C.

Chemical Weed Control in Lawns and Turf (continued)

Herbicide and Formulation	Brand Names	Amount of Formulation per 1,000 sq ft	Amount of Formulation per Acre	Pounds Active Ingredient per Acre	Precautions and Remarks
Preemergence and Postemergence Control, Annual Bluegrass (continued)					
ethofumesate, MOA 8 (4 SC)	(4 SC): Phoenix Thrasher, PoaConstrictor	0.55 to 1.47 fl oz	1.5 to 4 pt	0.75 to 2	Must be professionally applied to residential and nonresidential turf including golf courses and sod farms. May be applied to established perennial ryegrass, Kentucky bluegrass, creeping bentgrass, tall fescue, St. Augustinegrass, and dormant bermudagrass. Do not apply to putting greens. Delay application at least 8 weeks after a pgr application. Fall annual bluegrass control best during period of maximum germination. Spring applications most effective following fall applications. For overseeded bermudagrass, apply 1 to 2 weeks after perennial ryegrass emergence and repeat at 21- to 28-day intervals. Do not apply to bermudagrass 4 weeks prior to breaking winter dormancy.
imazaquin + prodiamine + simazine (5 L)	(2.5 G) Coastal	1.1 to 1.47 oz	48 to 64 oz	1.88 to 2.5	For use on established bermudagrass, centipedegrass, St Augustinegrass, and Zoysiagrass. Approved application sites are golf courses (excluding putting greens), athletic fields, commercial and residential turf, and sod farms. Should be applied from 15 September to 31 May for preemergent and early postemergent control of annual bluegrass when applied in fall. Will also provide preemergent control of summer annual grasses such as crabgrass if applied prior to grass germination in late winter/early spring. Also provides control of various broadleaf weeds. See label for additional precautions and weeds controlled.
Postemergence Control and Seedhead Suppression, Annual Bluegrass in Overseeded Bermudagrass Fairways, Tees					
bispyribac-sodium, MOA 2 (17.6 SG)	Velocity	0.046 to 0.138 oz	2 to 6 oz	0.021875 to 0.065625	Do not apply to putting greens, ryegrass mowed to less than 0.375 inch, or non-overseeded bermudagrass. Apply between Feb. 1 and March 15. Make first application when annual bluegrass begins flowering. If actively flowering, use the low rate and re-treat in 28 to 35 days. If not actively flowering, use the low rate and retreat in 14 to 21 days with the low rate. Do not apply if air temperature is less than 50 degrees F within 3 days after application. Check label for further special instructions.
amicarbazone, MOA 5 (70 WG)	Xonerate	0.023 to 0.23 oz	1 to 10 oz	0.044 to 0.44	Also tolerant to 6 month established turfgrasses such as bahiagrass, centipedegrass, seashore paspalum, St. Augustinegrass, zoysiagrass, bentgrass, Kentucky bluegrass, perennial ryegrass, fine and tall fescue. Labeled for use on golf course, sod farm, residential, commercial, athletic field and roadside turf. Bentgrass tees: 1 ounce/acre at 7 day intervals for 4 applications. Bentgrass roughs and fairways: 2 to 3 ounces/acre for 14 to 21 day intervals for 2 applications. Cool season turf: 2 to 4 ounces/acre for 14 to 21 day intervals for 2 applications. Warm season turf: 3 to 10 ounces/acre for 14 to 21 day intervals for 2 applications not to exceed 10 ounces/acre per year. Allow 4 weeks before cutting or lifting sod. Allow 1 week before overseeding winter grasses.
Postemergence Control, Annual Bluegrass, Overseeded Perennial Ryegrass, Tall Fescue, *Poa trivialis*					
flazasulfuron, MOA 2 (25 DG)	Katana	0.011 to 0.069 oz	0.5 to 3 oz	0.0078 to 0.0469	For use on well established bermudagrass, zoysiagrass, centipedegrass, and seashore paspalum grown turf including golf courses (including fairways, roughs, greens (bermudagrass and seashore paspalum only), tees, collars and approaches), industrial parks, tank-sod- and seed farms, cemeteries, athletic field and commercial lawns. Residential turf applications are limited to spot applications. Apply a maximum of 1.5 ounces per acre on fully green centipedegrass and seashore paspalum. 3 ounces per acre needed for annual bluegrass control and best if applied in spring. 0.5 to 1.5 ounces per acre will control perennial and Italian ryegrass. For clumpy ryegrass, use 1.5 to 3 ounces per acre. 1.5 ounces per acre needed for tall fescue control. 2.25 to 3 ounces per acre needed for *poa trivialis* control. Include a nonionic surfactant at 0.25% by volume.
foramsulfuron, MOA 2 (0.19 SC)	Revolver	0.2 to 0.6 fl oz	8.8 to 26.2 fl oz	0.013 to 0.039	For use on bermudagrass and zoysiagrass grown on home lawns, golf courses and sod farms. Do not use on warm season turfgrass collars surrounding bentgrass greens. May be applied up to 1 week prior to overseeding. Do not apply within 2 weeks of bermudagrass sprigging. Apply in 25 to 60 gallons water per acre. Rainfast after 2 hours. Surfactant not required.
rimsulfuron, MOA 2 (25 DF)	Tranzxit GTA, Rimsulfuron	0.011 to 0.092 oz	0.5 to 4 oz	0.0078 to 0.0625	May be applied to bermudagrass, zoysiagrass and centipedegrass on professionally managed sports facilities at professional and collegiate levels, golf courses, sod farms, roadsides, industrial and commercial lawns. For annual bluegrass control, apply November through December and again February through March if needed at 2 ounces per acre. May be applied 10 to 14 days prior to overseeding. For overseeded removal, apply 2 ounces per acre 3 to 4 weeks before desired removal date, and repeat 3 weeks later if needed. For weed control along roadsides, apply 4 ounces per acre if single application only. A nonionic surfactant at 0.25% by volume or an oil adjuvant such as crop oil concentrate and modified seed oil at 1% by volume are required. Do not apply to cool-season turfgrasses, residential lawns or newly sprigged/sodded bermudagrass.

Chemical Weed Control in Lawns and Turf (continued)

Herbicide and Formulation	Brand Names	Amount of Formulation per 1,000 sq ft	Amount of Formulation per Acre	Pounds Active Ingredient per Acre	Precautions and Remarks
Postemergence Control, Annual Bluegrass, Overseeded Perennial Ryegrass, Tall Fescue, Poa trivialis					
[metsulfuron + rimsulfuron], MOA 2 + 2 (37 WG)	Negate	0.0344352 oz	1.5 oz	0.0346875	Use on well established bermudagrass and zoysiagrass grown on nonresidential turf including golf courses, sod farms, industrial and commercial lawns, and professionally managed college and professional sports fields. Overseeding can occur 2 months after application. Include a nonionic surfactant at 0.25% by volume.
sulfosulfuron, MOA 2 (75 DG)	Certainty, Outrider	0.017 to 0.046 oz	0.75 to 2 oz	0.035 to 0.09375	May be applied to certain ornamental native grasses and also bermudagrass species, zoysiagrass, centipedegrass, St. Augustinegrass, and kikuyugrass grown on sod farms, golf courses (excluding greens), commercial and residential turf that is highly managed, and other noncrop areas. Use 1.5 to 2 ounces per acre for fall annual bluegrass control 7 to 10 days before overseeding. Use 0.75 to 1.25 ounces per acre for fall or winter control in nonoverseeded bermudagrass, and reapply if needed but not before 21 days after initial application. For tall fescue control, two applications may be required at 4- to 10-week intervals. Perennial ryegrass control not as complete as with foramsulfuron, rimsulfuron, or trifloxysulfuron. Use a nonionic surfactant at 0.25% by volume. Do not exceed 2.66 ounces per acre per year.
trifloxysulfuron, MOA 2 (75 WG)	Monument	0.0023 to 0.0129 oz	0.1 to 0.56 oz	0.0047 to 0.0263	May be applied to residential bermudagrass and zoysiagrass and also on golf courses, sod farms, and other nonresidential turf areas. A nonionic surfactant at 0.25 to 0.5% by volume is recommended. Temporary discoloration may occur if used with MSO or COC. May be applied 3 weeks prior to overseeding. Use rates of 0.1 to 0.3 ounces per acre to remove overseeded perennial ryegrass and *Poa trivialis* to aid bermudagrass spring transition. Labeled turf species can be seeded or sprigged into treated areas 4 weeks after application.
Preemergence and Postemergence Control, Annual Bluegrass and certain winter annual broadleaf weeds					
atrazine, MOA 5 (4 L)	(4 L): AAtrex 4 L, Atrazine 4 L	0.75 to 1.5 fl oz	1 to 2 qt	1 to 2	Use on centipedegrass, St. Augustinegrass, and dormant bermudagrass. Apply Nov. 15 to Dec. 31. Follow label directions.
(90 DF, 90 WG)	(90 DF, 90 WDG): AatrexNine-O, Atrazine 90 WDG, Atrazine 90 DF	0.025 to 0.05 lb	1.1 to 2.2 lb		
simazine, MOA 5 (90 WDG, 90 DF)	(90 DF, 90 WDG): Simazine, Sim-Trol, Princep Caliber 90	0.4 to 0.8 oz	1.1 to 2.2 lb	1 to 2	Use on bermudagrass, centipedegrass, St. Augustinegrass, and zoysiagrass. See label for instructions on newly sprigged turfgrass or on hybrid bermudagrass. Apply Nov. 15 to Dec. 15. Follow label directions.
(4 L)	(4 L): Princep Liquid, Simazine, Sim-Trol	0.75 to 1.5 fl oz	1 to 2 qt		
Preemergence Control, Certain Broadleaf Weeds					
isoxaben, MOA 21 (75 DF, 75 WG)	Gallery, Isoxaben	0.25 to 0.5 oz	0.66 to 1.33 lb	0.5 to 1	All established turfgrasses are tolerant. However, do not apply to putting greens or turfgrass grown for seed. Check label for specific weeds controlled.
pendimethalin, MOA 3 (3.8 CS)	(3.8 CS): HydroCap, Pendulum AquaCap, Pre-M AquaCap, Prowl H2O, Satellite HydroCap	1.15 to 1.55 fl oz	3.1 to 4.2 pt	1.5 to 2	See section on preemergence control of crabgrass or product label for turfgrass tolerance. Provides preemergence control of summer broadleaf weeds, such as prostrate spurge, prostrate knotweed, and purslane species, as well as winter broadleaf weeds, such as yellow woodsorrel, hop clover, cudweed species, common chickweed, lawn burweed, henbit, and corn speedwell when applied before expected germination.
Preemergence Control of Smooth and Large Crabgrass, Goosegrass, Other Broadleaf Weeds					
[isoxaben + dithiopyr]	(0.75 GR) Fortress	3.4 to 4.6 lb	140 to 200 lb	1.13 to 1.5	Use on established turfgrasses (creeping bentgrass, Kentucky bluegrass, fine fescue, tall fescue, perennial ryegrass, bahiagrass, bermudagrass, carpetgrass, centipedegrass, seashore paspalum, St Augustinegrass, and zoysiagrass). Do not use on putting greens. Use for preemergent control of annual grasses and broadleaf weeds. See label for a complete list of weeds controlled. Also has postemergent control of small crabgrass (pre-tiller).
Preemergence and Postemergence Control Crabgrass, Goosegrass, Other Annual Grasses, Broadleaf Weeds, Sedges					
mesotrione, MOA 27 (4 SC)	Tenacity	0.092 to 0.183 fl oz	4 to 8 fl oz	0.125 to 0.25	Use on residential turf, golf courses (not greens) and sod farms for pre- and postemergence weed control. Tolerant turfgrasses include St. Augustinegrass, centipedegrass, tall fescue, fine fescue, Kentucky bluegrass, and perennial ryegrass. Add a nonionic surfactant and repeat application after 2 to 3 weeks for improved postemergence control. Tank mix with prodiamine 65 WG for extended preemergence grassy weed control. Can be applied at seeding to all tolerant grasses except fine fescue. After turf germination, wait 4 weeks or until turf has been mowed twice before making a postemergence application. Also controls henbit, chickweed, dandelion, white clover, Florida betony, Florida pusley, ground ivy, oxalis, wild violet, creeping bentgrass, and yellow nutsedge.

Chemical Weed Control in Lawns and Turf (continued)

Herbicide and Formulation	Brand Names	Amount of Formulation per 1,000 sq ft	Amount of Formulation per Acre	Pounds Active Ingredient per Acre	Precautions and Remarks
Preemergence and Postemergence Control Crabgrass, Goosegrass, Other Annual Grasses, Broadleaf Weeds, Sedges					
[sulfentrazone + prodiamine], MOA 14 + 3 (4 SC)	Echelon	0.184 to 0.826 fl oz	0.5 to 2.25 pt	0.25 to 1.125	For use in residential and institutional lawns, athletic fields, sod farms, golf course fairways and roughs, roadsides, utility rights-of-way, railways, and industrial areas. Apply to turf following a second mowing if a good root system has been established. Apply up to 12 fluid ounces per acre to bentgrass at 0.5 inch or higher, fine fescue, and perennial ryegrass. Apply 18 to 24 fluid ounces per acre to perennial bluegrass, tall fescue, and all warm season grasses except St. Augustinegrass (do not apply) and bermudagrass (apply 18 to 36 fluid ounces per acre). For sod production, apply 6 months after establishment, and do not harvest within 3 months. Do not apply with adjuvants or surfactants. [Sulfentrazone + prodiamine should not be applied to cool-season turf with N-containing fertilizers unless some short-term discoloration is tolerable.
Postemergence Control, Crabgrass, Goosegrass					
fenoxaprop, MOA 1 (0.57 EC)	Acclaim Extra	0.3 to 0.9 fl oz	0.8 to 2.4 pt	0.057 to 0.174	Use only on perennial ryegrass, fine fescue, tall fescue, Kentucky bluegrass, and zoysiagrass. Reduced vigor or discoloration can occur. Rate depends upon leaf number or tillers of grass weeds and turf tolerance. Check label. A second application may be applied after 14 days.
		0.08 fl oz	3.5 fl oz	0.016	Apply only to established Penncross bentgrass maintained at a minimum cutting height of at least 0.25 inch. Bentgrass should be established for one growing season. Do not apply to greens. Applications should be made at a minimum of 21-day intervals, beginning in the spring when grassy weeds first emerge and are not larger than two-leaf. Repeat applications throughout the summer as new infestations of one- to two-leaf grassy weeds occur. See label for other restrictions.
metribuzin, MOA 5 (75 DF)	(75 DF): Dimetric DF, Glory, Metribuzin 75, Tricor DF	0.12 to 0.24 oz	0.33 to 0.67 lb	0.25 to 0.5	Recommended for application by commercial applicators only on established bermudagrass turf (such as parks, athletic fields, golf course fairways, cemeteries, and sod farms) that has a mowing height of 0.5 inch or greater. Apply when turf is vigorously growing and not under stress. Repeat if necessary in 7 to 10 days. Do not make more than two applications per season. Do not apply to greens, tees, or aprons.
sethoxydim, MOA 1 (1 EC)	(1 EC): Poast Plus, Segment, Sethoxydim SPC	0.8 to 1.38 fl oz	2.25 to 3.75 pt	0.28 to 0.47	Use in seedling and established centipedegrass and fine fescues. Apply 2.25 pint to grasses up to 6 inches and 3.75 pints to grasses up to 12 inches if turf is tolerant. Does not control yellow and purple nutsedge, annual bluegrass or broadleaf weeds. Apply no sooner than 3 weeks after spring greenup of centipedegrass. Apply before crabgrass becomes extensively tillered. Delay all treatments until newly planted centipedegrass has 3 inches of new stolon growth. Do not mow within 7 days before or after application. Two applications 3 weeks apart will suppress bahiagrass. Additives or adjuvants not required.
Postemergence Control, Smooth and Large Crabgrass, Barnyardgrass, White and Hop Clover, Common Dandelion, Dollarweed, Foxtails					
quinclorac, MOA (27 + 4) (75 DF)	(75 DF): Armortech Quinclorac Pro, Quinclorac, Quinclorac SPC, Quinstar	0.367 oz	1 lb	0.75	For use in residential and nonresidential turf that is established or newly seeded, overseeded, or sprigged. Refer to label for specific varieties. Apply to common and hybrid bermudagrass, Kentucky bluegrass, annual bluegrass, buffalograss, tall fescue, annual and perennial ryegrass, creeping bentgrass, and zoysiagrass. Can also be applied to fine fescue but must be in a blend. Some discoloration of hybrid bermudagrass, creeping bentgrass or fine fescue may occur. Do not apply to bahiagrass, centipedegrass, St. Augustinegrass, or dichondra. Do not use on golf course greens or collars. The addition of methylated seed oil (1.5 pints per acre or 0.55 ounces per 1,000 square feet) or a crop oil concentrate (2 pint per acre or 0.73 ounces per 1,000 square feet) is required for control. Application to weeds under stress will result in poor control. Irrigation 24 hours prior to application is recommended if drought conditions exist. Some ornamental plants are sensitive to quinclorac. See label for further precautions.
(1.5 SL)	(1.5 SL): Drive XLR8, Quinclorac 1.5 L	1.45 fl oz	2 qt		
Postemergence Control, Smooth and Large Crabgrass, Barnyardgrass, Foxtails, and many broadleaf weeds					
[quinclorac + sulfentrazone + 2,4-D amine + dicamba], MOA (27 + 4) + 14 + 4 + 4 (1.79 L)	Q4 Plus	1.8 to 3 fl oz	5 to 8 pt	1.12 to 1.79	For use in fully dormant bermudagrass as well as actively growing bermudagrass after spring greenup but use only 5 to 7 pints per acre. Also labeled in fully dormant zoysiagrass s well as cool-season turf including annual bluegrass and ryegrass, perennial bluegrass and ryegrass, and fescue species. Do not apply to bahiagrass, bentgrass (creeping, Seaside, Colonial), centipedegrass, St. Augustinegrass, carpetgrass, and golf course greens, tees, and collars. May be applied to home lawns. Apply to seedling grasses after second or third mowing, or 28 days after emergence. Wait 3 to 4 weeks after sodding, sprigging, or plugging operations to apply. Wait 4 weeks after application to seed.
[quinclorac + mecoprop + dicamba], MOA (27 + 4) + 4 + 4 (2.45 SL)	Onetime	0.5 to 1.45 fl oz	0.68 to 2 qt	0.4165 to 1.225	For use in warm- and cool-season residential and non-residential turf, including but not limited to commercial property, parks, roadsides, schools, athletic fields, cemeteries, and golf courses. May be applied to species of bermudagrass, bluegrass, fescue, and ryegrass as well as creeping bentgrass, seashore paspalum, and zoysiagrass. Use with methylated seed oil at 1.5 pints per acre. Allow 28 days of seedling or sprig growth before application. If treating first, allow 28 days before seeding or sprigging. Do not apply to golf course collars or greens or to turf grown for sod. Use low rate in 2 split applications when treating creeping bentgrass.

Chemical Weed Control in Lawns and Turf (continued)

Herbicide and Formulation	Brand Names	Amount of Formulation per 1,000 sq ft	Amount of Formulation per Acre	Pounds Active Ingredient per Acre	Precautions and Remarks
Postemergence Control, Smooth and Large Crabgrass, Barnyardgrass, Foxtails, and many broadleaf weeds (continued)					
[carfentrazone + quinclorac], MOA 14 + (27 + 4) (75 WG)	SquareOne	0.184 to 0.413 oz	8 to 18 oz	0.35 to 0.79	Can use up to 12 ounces per acre 7 days after emergence from seed or sod installment on bluegrass and fescue species and perennial ryegrass; 18 ounces per acre can be used 7 days after seed, sod or sprig operations on bermudagrass species, centipedegrass and seashore paspalum. Wait 14 days after emergence for zoysiagrass. May apply to residential, commercial, and institutional lawns, athletic fields, sod farms, and golf course fairways and roughs. Adjuvants not required but may help on mature weeds.
[sulfentrazone + quinclorac], MOA 14 + (27 + 4) (75 WG)	Solitare	0.367 to 0.735 oz	1 to 2 lb	0.75 to 1.5	Use up to 21 ounces per acre on well-established tall fescue, Kentucky bluegrass and perennial ryegrass; up to 32 ounces per acre on well-established bermudagrass, centipedegrass, zoysiagrass and seashore paspalum. May be applied to residential, commercial, and institutional lawns, athletic fields, sod farms, and golf course fairways and roughs. After treatment, wait at least 1 month before reseeding, overseeding (use slit seeder for best results), or sprigging. Wait at least 3 months for sod establishment and do not spray within 3 months of harvest. Controls goosegrass in the 1 to 4 leaf stage. Yellow nutsedge and kyllinga species are also controlled. Do not apply with a spray adjuvant.
[fenoxaprop + fluroxypyr + dicamba], MOA 1 + 4 + 4 (0.75 EC)	Last Call	1.3 to 1.5 fl oz	3.5 to 4 pt	0.33 to 0.375	Tolerant turfgrass species include zoysiagrass, Kentucky bluegrass, perennial ryegrass, fine and tall fescue. May be applied to golf courses excluding greens and tees, athletic fields, commercial and residential turf. Sod farm use is not permitted. Best grass weed control will be achieved when treated from 1 leaf to 4 tiller stage. Do not apply more than 15 pints per acre per year. Do not reapply within 14 days of an application. Surfactant not required. Spot treat using 0.6 to 1 fluid ounces per 1 gallon water.
Postemergence Control, Large Crabgrass, Carpetgrass, Bull Paspalum, Bahiagrass, Foxtails, and many broadleaf weeds, including Chamberbitter, Corn Speedwell, Dichondra, Dollarweed, Doveweed, Florida Betony, Florida Pusley, Lespedeza, Oxalis, Spurge, Virginia Buttonweed, Kyllinga					
[thiencarbazone-methyl + iodosulfuron-methyl + dicamba], MOA 14 + 2 + 4 (68 WG)	Celsius WG	0.057 to 0.113 oz	2.5 to 4.9 oz	0.106 to 0.208	For use by licensed applicators in residential and commercial lawns, golf courses (excluding greens), sports fields, parks, recreational areas, roadsides, school grounds, and sod farms. Provides up to 60 days residual control. Use on bermudagrass, zoysiagrass, centipedegrass, and St Augustinegrass. Apply maximum 7.4 ounces per acre per season. Safe to use at high temperatures. Ryegrass can be overseeded 2 weeks after application. Apply 30 days prior to seeding bermudagrass or zoysiagrass. Wait 2 weeks after bermudagrass seedling emergence or sprigging operation before applying. For zoysiagrass, wait 3 weeks after seedling emergence before applying. A nonionic surfactant or methylated seed oil at 0.25% v/v is required for optimum control.
Postemergence Control or Suppression of Summer Weeds Such as Crabgrass Species, Goosegrass, Dallisgrass, Virginia Buttonweed, Doveweed, Florida Pusley, Nutsedge and Kyllinga Species; Winter Weeds Such as Poa annua, Poa trivialis, Tall Fescue, Henbit, Corn Speedwell, and Species of Ryegrass, Chickweed, and Clover					
[thiencarbazone-methyl + foramsulfuron + halosulfuron], MOA 14 + 2 + 2 (60.5 WG)	Tribute Total	0.0735 oz	3.2 oz	0.121	Apply to well-established residential and commercial bermudagrass and zoysiagrass (Emerald, Meyer, Zeon) lawns, golf courses (excluding greens), athletic fields, sod farms, roadsides, parks, cemeteries and recreational areas. Do not exceed 3.2 ounces per acre per application or 6.4 ounces per acre yearly. Use 0.25 to 0.5% by volume nonionic surfactant or 0.5 to 1% by volume methylated seed oil. After application, wait 12 weeks to overseed ryegrass or bermudagrass. Wait 1 month after bermudagrass seedling emergence and 2 weeks after sprigging or sodding bermudagrass before treating. Temporary stunting and yellowing may last up to 2 weeks, but turf will recover. Crabgrass and goosegrass are controlled up to 2 tiller stage.
Postemergence Control, Goosegrass					
[(2,4-D + MCPP + dicamba + carfentrazone) + topramezone]		1.5 + 0.006 oz	4 pts + 0.25 oz	1.1 + 0.0005	Apply to established bermudagrass and emerged goosegrass. Will control mature goosegrass but better control obtained when applied to smaller goosegrass. Bermudagrass discoloration will occur and typically lasts less than 2 weeks. Mixing the products vastly reduces whitening on bermudagrass from topramezone. More discoloration will occur when temperatures are in excess of 85 degrees. Do not apply to putting greens.
foramsulfuron, MOA 2 (0.19 SC)	Revolver	0.39 fl oz	17 fl oz	0.025	For use on bermudagrass and zoysiagrass grown on home lawns, golf courses and sod farms. See precautions listed under annual bluegrass section. For goosegrass control, apply 17 fl ounces per acre on plants up to 2 tillers followed by 17 fluid ounces per acre 2 weeks later.
metribuzin, MOA 5 (75 DF)	Sencor	0.18 oz	8 oz	0.38	Apply to established bermudagrass and emerged goosegrass. Will control mature goosegrass but better control obtained when applied to smaller goosegrass. Irrigate in immediately with 0.25 inches of water. Do not apply to saturated soils or if significant rainfall is expected. Immediate irrigation increases efficacy and reduces bermudagrass discoloration. If watered in immediately, limited discoloration will occur. Do not apply to putting greens.
[metribuzin + topramezone]	Sencor + Pylex	0.09 oz + 0.006 oz	4 oz + 0.25 oz	0.19 + 0.005	Apply to established bermudagrass and emerged goosegrass. Will control mature goosegrass but better control obtained when applied to smaller goosegrass. Do not irrigate in. Some slight bermudagrass discoloration will occur but disappears in approximately 10 to 14 days. Mixing the products vastly reduces whitening on bermudagrass from topramezone. Do not apply to putting greens.

Chemical Weed Control in Lawns and Turf (continued)

Herbicide and Formulation	Brand Names	Amount of Formulation per 1,000 sq ft	Amount of Formulation per Acre	Pounds Active Ingredient per Acre	Precautions and Remarks
Postemergence Control, Goosegrass (continued)					
sulfentrazone, MOA 14 (4 SC)	Dismiss Turf	0.275 fl oz	0.75 pt	0.375	May be applied to home lawns. For use on creeping bentgrass, tall and fine fescue, perennial ryegrass, Kentucky bluegrass, and all warm-season turf species except St. Augustinegrass. See precautions listed under purple and yellow nutsedge section. For goosegrass control, apply 0.75 pint per acre on plants up to 2 tillers.
Postemergence Control, Sedge and Various Broadleaf Weeds					
[sulfentrazone + imazaquin]	Surepyc IQ	0.5 to 1 oz	22 to 44 oz	0.38 to 0.75	Use on established bermudagrass (common and hybrid), centipedegrass, St. Augustinegrass, and zoysiagrass. Do not use on putting greens. See label for further restrictions. Controls kyllinga, yellow and purple nutsedge, dandelion, henbit, lawn burweed, spurge, wild garlic, yellow woodsorrel. See label for additional weeds controlled. Addition of a nonionic surfactant at 0.25% volume/volume is required.
Postemergence Control, Bahiagrass, Crabgrass, Dallisgrass, Goosegrass, Nutsedge, Annual Sedges, Sandbur					
MSMA, MOA 17 (6 SL, 6.6 SL)	(6 SL): Target 6 Plus, Drexel MSMA 6 Plus (6.6 SL): Target 6.6 Plus, Drexel MSMA 6.6 Plus		several concentrations	1.82 to 4.5	MSMA is only registered for golf course, sod farm, and highway right-of-way use. Bermudagrass, bluegrass and zoysiagrass are tolerant. Injury may result on bentgrass, fescue and also St. Augustinegrass grown for commercial sod production only. Do not use on carpetgrass or centipedegrass. MSMA restrictions: For existing golf courses, spot treat (100 square feet per spot) not to exceed 25% of total acreage. For new courses, make 1 broadcast application per year. For sod farms, make 1 to 2 broadcast applications per year and maintain 25 feet buffer around permanent water bodies. For highway rights of way, make 2 broadcast applications and maintain 100 feet buffer around permanent water bodies.
Postemergence Control, Crabgrass, Goosegrass, Sandbur, Dallisgrass					
MSMA, MOA 17 (6 SL, 6.6 SL)	(6 SL): Target 6 Plus, Drexel MSMA 6 Plus (6.6 SL): Target 6.6 Plus, Drexel MSMA 6.6 Plus		several concentrations	1.5 to 2	See remarks for MSMA and metribuzin. The combination improves goosegrass control. Should be applied to bermudagrass only.
+ metribuzin, MOA 5 (75 DF)	Sencor		+ 0.17 to 0.33 lb	+ 0.125 to 0.25	
Postemergence Control, Crabgrass, Goosegrass, Sandbur					
asulam, MOA 18 (3.34 SL)	Asulox	1.8 fl oz	5 pt	2	Use only on St. Augustinegrass and Tifway 419 turf. On golf courses, use only on fairways and roughs.
Postemergence Control, Crabgrass and Foxtail Species, Goosegrass, Broadleaf Signalgrass, Japanese Stilgrass **Postemergence Suppression, Creeping Bentgrass, Common Bermudagrass, Dallisgrass, Nimblewill**					
topramezone, MOA 27 (2.8 L)	Pylex	0.023 to 0.034 fl oz	1 to 1.5 fl oz	0.021875 to 0.0328125	Labeled for broadcast treatment use in residential and athletic field turf, as well as in nonresidential turf sites including sod farms, golf courses (excluding greens and collars), parks, roadsides, cemeteries, and commercial properties. Tolerant turf species include Kentucky bluegrass, tall and fine fescue, perennial ryegrass, and centipedegrass at seeding and then anytime beyond 28 days after seeding. Add crop oil concentrate or methylated seed oil for enhanced control at 0.5 to 1% by volume. Don't apply greater than 2 fluid ounces per acre per application or 4 fluid ounces per acre per year. Bleaching intensity of susceptible weeds reduced and broadleaf weed spectrum increased if tankmixed with quinclorac, [quinclorac + mecoprop + dicamba] or triclopyr. For suppression of above-listed weeds, add triclopyr at 1 pound ai per acre and make either 2 or 3 applications at 3 to 4 week intervals depending on topramezone rate. Creeping bentgrass is marginally tolerant to topramezone at 0.25 fluid ounces per acre. Test on a small area before large-scale use. Sequential applications may be required to achieve desired level of weed control.
Postemergence Control, Yellow Nutsedge, Annual Sedge					
bentazon, MOA 6 (4 SL)	Basagran Sedge Control, Basagran T/O, Bentazon 4, Lescogran	0.75 to 1.5 fl oz	1 to 2 qt	1 to 2	For control of yellow nutsedge in established bluegrass, fescues, bentgrass, ryegrass, bermudagrass, bahiagrass, St. Augustinegrass, centipedegrass, and zoysiagrass. Apply to yellow nutsedge when actively growing under good soil moisture conditions. Additional applications may be made at intervals of 10 to 14 days until nutsedge is controlled.
Postemergence Control, Purple and Yellow Nutsedge, Kyllinga Species					
flazasulfuron, MOA 2 (25 DG)	Katana	0.034 to 0.069 oz	1.5 to 3 oz	0.023 to 0.0469	For use on well-established bermudagrass, zoysiagrass, centipedegrass and seashore paspalum grown on nonresidential turf including golf course fairways, roughs and tees, and industrial parks, tank-sod- and seed farms, cemeteries, athletic field and commercial lawns. Apply a maximum of 1.5 ounces per acre on fully green centipedegrass and seashore paspalum. 3 ounces per acre needed for perennial nutsedge and some annual sedge species control. Repeat applications in 2 to 6 weeks when nutsedge or sedge growth is evident. 1.5 to 2.25 ounces per acre will control kyllinga species. Maintain a 25 feet nontreated border beside susceptible turf species. Can overseed in 2 weeks if applied up to 1.5 ounces per acre. Wait 4 weeks if applied more than 1.5 ounces per acre. Include a nonionic surfactant at 0.25% by volume.

Chemical Weed Control in Lawns and Turf (continued)

Herbicide and Formulation	Brand Names	Amount of Formulation per 1,000 sq ft	Amount of Formulation per Acre	Pounds Active Ingredient per Acre	Precautions and Remarks
Postemergence Control, Purple and Yellow Nutsedge, Kyllinga Species (continued)					
imazaquin, MOA 2 (70 DG)	Image 70 DG	0.128 to 0.256 oz	0.357 to 0.714 lb	0.25 to 0.5	Use on bermudagrass, centipedegrass, St. Augustinegrass, and zoysiagrass. Do not apply during spring greenup. Temporary yellowing may occur. Add a nonionic surfactant at 2 pt per 100 gal of spray solution. Addition of MSMA at 1.5 lb active per acre will improve sedge control in MSMA tolerant turfgrasses.
imazosulfuron, MOA 2 (75 WG)	Celero	0.184 to 0.322 oz	8 to 14 oz	0.38 to 0.66	May be applied to established (two mowings) residential and commercial bermudagrass, zoysiagrass, centipedegrass, St. Augustinegrass, creeping bentgrass, Kentucky bluegrass, perennial ryegrass, tall fescue, and fine fescue. Do not apply to putting greens. Reapply 3 weeks after initial application when using the 8 ounces per acre rate. Reapply as needed 3 weeks after initial application when using rates above 8 ounces per acre. Wait 4 weeks to seed or sod after application. Use an 80% active nonionic surfactant at 0.25% by volume. For spot treatment, add 0.25 to 0.33 oz in 1 to 2 gallons of water per 1000 square feet. Add 2 teaspoons nonionic surfactant per gallon.
halosulfuron, MOA 2 (75 WDG)	HiYield Nutsedge Control, Nutgrass Killer II, Profine 75, Sandea	0.9 g	0.67 to 1.33 oz	0.031 to 0.062	May be applied to established residential and commercial bermudagrass, bahiagrass, zoysiagrass, centipedegrass, St. Augustinegrass, creeping bentgrass, Kentucky bluegrass, perennial ryegrass, tall fescue, and fine fescue. Apply broadcast when sedges have reached the 3- to 8-leaf stage. Use lower rate for light infestations and higher rate for heavy infestations. A second treatment will usually be required 6 to 10 weeks after the initial treatment. Use an 80% active nonionic surfactant at 2 quarts per 100 gallons of spray solution (0.5% by volume). Do not exceed 1 to 2 pints of surfactant per acre. Do not apply to putting greens. Halosulfuron only suppresses green kyllinga.
MSMA, MOA 17 (6 SL, 6.6 SL)	(6 SL): Target 6 Plus, Drexel MSMA 6 Plus (6.6 SL): Target 6.6 Plus, Drexel MSMA 6.6 Plus		several concentrations	2 to 3	See remarks for MSMA above. Will require at least 2 applications 7 to 10 days apart.
sulfosulfuron, MOA 2 (75 DG)	Certainty, Outrider	0.017 to 0.029 oz	0.75 to 1.25 oz	0.035 to 0.059	May be applied to certain ornamental native grasses and also bermudagrass species, zoysiagrass, centipedegrass, St. Augustinegrass, and kikuyugrass grown on sod farms, golf courses (excluding greens), commercial and residential turf that is highly managed, and other noncrop areas. Use 0.75 to 1.25 ounces per acre, and repeat in 4 to 10 weeks if needed. Use a nonionic surfactant at 0.25% by volume.
trifloxysulfuron, MOA 2 (75 WG)	Monument	0.0023 to 0.0129 oz	0.1 to 0.56 oz	0.0047 to 0.0263	May be applied to residential bermudagrass and zoysiagrass and also on golf courses, sod farms, and other nonresidential turf areas. A nonionic surfactant at 0.25 to 0.5% by volume is recommended. Temporary discoloration may occur if used with MSO or COC. Use rates of 0.33 to 0.56 ounces per acre for sedge and kyllinga species control. Labeled turf species can be seeded or sprigged into treated areas 4 weeks after application. Repeat application may be needed in 4 to 6 weeks.
Postemergence Control, Purple and Yellow Nutsedge, Kyllinga Species, and various broadleaf weeds					
sulfentrazone, MOA 14 (4 SC)	Dismiss Turf	0.092 to 0.275 fl oz	0.25 to 0.75 pt	0.125 to 0.375	May be applied to home lawns. For use on creeping bentgrass, tall and fine fescue, perennial ryegrass, Kentucky bluegrass, and all warm-season turf species except St. Augustinegrass. Wait 3 months to seed, overseed, or sprig unless overseeding bermudagrass with perennial ryegrass, which only requires a 4- to 6-week waiting period after application. Apply to seedling grasses after second mowing and to new sod 6 months after establishment.
[sulfentrazone + imazethapyr], MOA 14 + 2 (4 SC)	Dismiss South	0.22 to 0.33 fl oz	9.5 to 14.4 fl oz	0.29 to 0.45	May be applied to home lawns, athletic fields, sod farms, golf course fairways and roughs, and various non-crop sites. For use on bahiagrass, bermudagrass, centipedegrass, and zoysiagrass. Do not apply to soils classified as sand with less than 1% organic matter. Do not reseed, overseed, or sprig within 1 month of application. Expect slight perennial ryegrass injury if overseeded 2 to 4 weeks after application. Allow 3 month sod establishment before treatment.
[sulfentrazone + metsulfuron], MOA 14 + 2 (66 WG)	Blindside	0.075 to 0.23 oz	3.25 to 10 oz	0.134 to 0.413	May be applied to established residential, commercial and institutional lawns, athletic fields, sod farms, and golf course fairways and roughs. Use up to 6.5 ounces per acre on Kentucky bluegrass and tall fescue and 10 ounces per acre on bermudagrass, centipedegrass, St. Augustinegrass and zoysiagrass. Do not reseed, overseed, or sprig within 1 month of application. Expect slight perennial ryegrass injury if overseeded 6 to 8 weeks after application. Allow 3 months sod establishment before treatment. No adjuvant needed.
Postemergence Control, Bahiagrass, Crabgrass, Yellow and Purple Nutsedge, Annual Sedge, Kyllinga Species					
imazapic, MOA 2 (2 AS)	Imazapic 2 SL, Impose, Panoramic, Plateau	0.092 to 0.184 fl oz	4 to 8 fl oz	0.063 to 0.125	For use on unimproved centipedegrass after complete greenup only. Not for use in home lawns. Do not use on other turfgrass species. A repeat application may be needed on tough to control perennial weeds such as bahiagrass. The highest labeled rate may discolor centipedegrass by causing a red color.
Postemergence Control, Dandelion, Carpetweed, Carolina Cranesbill, Curly Dock, Plantain, Dichondra, Shepherds-Purse, Yellow Rocket					
2,4-D amine, MOA 4 (4 SL)	(4 SL): various trade names	3 to 4 tsp	1.5 to 2 pt	0.75 to 1	Cut rate one-half for bentgrass, carpetgrass, centipedegrass, and St. Augustinegrass. Spray when weeds are young and actively growing. To reduce danger of injury to flowers and ornamentals by spray drift, use low pressure and do not spray on windy days.

Chemical Weed Control in Lawns and Turf (continued)

Herbicide and Formulation	Brand Names	Amount of Formulation per 1,000 sq ft	Amount of Formulation per Acre	Pounds Active Ingredient per Acre	Precautions and Remarks
Postemergence Control, Common Chickweed, Mouseear Chickweed, Creeping Charlie or Ground Ivy, Dandelion, Lespedeza, Black Medic, Spotted Spurge, Hop or White Clover					
mecoprop, MOA 4 (1.9 L) (1.16 L) (1.74 L)	MCPP-p4 Amine, Mecomec 2.5, Mecomec 4	1 to 1.5 fl oz 1.5 to 2.25 fl oz 0.75 to 1.5 fl oz	2.7 to 4 pt 4 to 6 pt 2 to 4 pt	0.64 to 0.95 0.58 to 0.87 0.43 to 0.87	Observe same precaution as for 2,4-D. May be used on bentgrass, carpetgrass, centipedegrass, St. Augustinegrass, and other turfgrasses.
Postemergence Control, Chickweed, White Clover, Dandelion, Curly Dock, Hawkweed, Henbit, Knotweed, Red Sorrel, Knawel, Spurweed, Spotted Spurge, Wild Strawberry, Yarrow					
dicamba, MOA 4 (4 SL)	Banvel, Topeka	1 to 2 tsp	0.5 to 1 pt	0.25 to 0.5	Apply as foliar spray to growing weeds. Prevent injury to ornamentals. Avoid rooting zone of shallow-rooted trees and shrubs.
diglycolamine, MOA 4 (4 SL)	Clarity, Clash, Detonate, Strut, Vanquish	1 to 4.5 tsp	0.5 to 2 pt	0.25 to 1	Do not exceed 1 pint per acre on bentgrass, carpetgrass, buffalograss, and St. Augustinegrass. Apply to newly seeded grasses after the second mowing. Do not exceed 0.25 pint per acre on extended sensitve plant roots on sandy soils and 0.5 pint per acre on clay soils.
Postemergence Control, All Weeds Listed Under 2,4-D Amine, MCPP, Dicamba, and Diglycolamine Sections					
[2,4-D amine + MCPP + dicamba], MOA 4 + 4 + 4 (various formulations)	Varios trade names	See individual label	See individual label	See individual label	Check individual labels for specific rates, instructions and precautions. Generally, 1) apply to grass seedlings after second mowing; 2) apply to sodded, sprigged, or plugged turf 3 to 4 weeks after operations; and 3) wait 3 to 4 weeks after application to seed. Many products labeled for tall fescue, perennial ryegrass, perennial bluegrass, bermudagrass, and St. Augustinegrass. Some products labeled for bentgrass putting greens, bahiagrass, zoysiagrass, and centipedegrass. Some products labled for home use when applied by a commercial applicator.
Postemergence Control, All Weeds Listed Under 2,4-D Amine, MCPP, Dicamba, and Diglycolamine Sections					
[2,4-D amine + MCPP + dichlorprop], MOA 4+ 4 + 4 (4.11 L) (2.48 L)	Spoiler, Triamine	0.62 to 1.47 fl oz 0.64 to 1.47 fl oz	1.7 to 4 pt 1.75 to 4 pt	0.873 to 2.06 0.543 to 1.24	
[MCPA + MCPP + dicamba], MOA 4+4+4 (4 L)	Ortho Weed B Gon Pro Southern, Tri-Power	0.7 to 1.5 fl oz	2.5 to 4.1 pt	1.25 to 2.05	
Postemergence Control, Curly Dock, Broadleaf Dock, Galinsoga, Nightshade, Clover (Red, Hop, White, Sweet), Goldenrod, Musk Thistle, Speedwells, Common Vetch, Hairy Buttercup, Broadleaf Plantain					
clopyralid, MOA 4 (3 EC)	Lontrel T&O	0.1 to 0.5 fl oz	0.25 to 1.33 pt	0.09 to 0.5	Do not apply to home lawns. May be used on bentgrass, Kentucky bluegrass, creeping, red, chewings, sheep and tall fescue, perennial ryegrass, bermudagrass, bahiagrass, buffalograss, centipedegrass, zoysiagrass, and St. Augustinegrass. Do not apply to putting greens and tees. Should be applied in a minimum of 20 gallons of water per acre. Surfactants are not necessary. Do not apply to exposed roots of certain trees and shrubs (legumes such as acacia, locust, mimosa, redbud, or mesquite) or *Tilia* spp. Do not use treated clippings for mulching and compost during the growing season of application.
All Weeds Listed Under 2,4-D Amine, Clopyralid, Dicamba, and Diglycolamine Sections					
[2,4-D amine + clopyralid + dicamba], MOA 4 + 4 + 4 (3.56 L)	Millennium Ultra 2	0.55 to 1.1 fl oz	1.5 to 3 pt	0.67 to 1.34	Do not apply to home lawns. Use on perennial bluegrass, ryegrass, and fescue species, bentgrass (excluding greens and tees), bermudagrass, zoysiagrass, and bahiagrass. Do not apply to seedling grasses until well established. Wait 3 to 4 weeks after application to seed.
Postemergence Control, Virginia Buttonweed, Chickweed Species, White Clover, Dandelion, Henbit, Ground Ivy, Prostrate Knotweed, Matchweed, Black Medic, Plantain Species, Common Woodsorrel					
[2,4-D amine + fluroxypyr + dicamba], MOA 4 + 4 + 4 (4 SL)	E-2, Escalade 2	0.36 to 1.1 fl oz	1 to 3 pt	0.5 to 1.5	Use on perennial bluegrass and ryegrass, tall fescue, creeping bentgrass (excluding greens and tees), bermudagrass species, bahiagrass, zoysiagrass, and St. Augustinegrass in residential, industrial, and institutional lawns, parks, cemeteries, athletic fields, golf courses, and sod farms. Use on St. Augustinegrass sod farms only. Apply 1 to 2 pints per acre on creeping bentgrass and 1.5 to 1.8 pints per acre on warm season turf grown for sod. Apply 2 to 3 pints per acre to all other turf areas. For non-turf areas, rate can be increased to 2 to 5 pinst per acre. Application can be made to grass seedlings after second mowing and to newly sodded, sprigged, or plugged grasses 3 to 4 weeks after operations.
[MCPA amine + fluroxypyr ester + dicamba], MOA 4 + 4 + 4 (4.8 SL)	Change Up	0.73 to 1.1 fl oz	2 to 3 pt	1.2 to 1.8	Same turf tolerances and uses as [2,4-D amine + fluroxypyr + dicamba] in addition to centipedegrass. Only spot treat St. Augustinegrass when temperature exceeds 80 degrees F. Do not apply more than two applications per year totaling 3 pints per acre. For non-turf areas, rate can be increased to 2 to 5 pints per acre. Application can be made to grass seedlings after second mowing and to newly sodded, sprigged, or plugged grasses 3 to 4 weeks after operations. Sod farm rates include 1.25 pints per acre for creeping bentgrass, 2 to 3 pints per acre for all other cool season grasses listed on label and 1.5 to 1.8 pints per acre for all warm season grasses listed on label.

Chemical Weed Control in Lawns and Turf (continued)

Herbicide and Formulation	Brand Names	Amount of Formulation per 1,000 sq ft	Amount of Formulation per Acre	Pounds Active Ingredient per Acre	Precautions and Remarks
Postemergence Control, Winter and Summer Annual Broadleaf Weeds					
bentazon + atrazine, MOA 6 + 5 Create by tank mixing				0.5 to 0.75 + 0.5 to 0.75	Apply to bermudagrass, centipedegrass, St. Augustinegrass, and zoysiagrass. Check individual labels for weeds controlled and weed size for proper application.
2,4-D (choline salt) + Fluroxypyr + halauxifen methyl (3.28 L)	GameOn	1.1 to 1.47 oz	48 to 64 oz	1.23 to 1.64	For postemergence control of annual and perennial broadleaf weeds in established turfgrass and ornamental grasses in golf courses, industrial sites, cemeteries, commercial sod farms, and unimproved turfgrass areas. Not for use in residential turfgrasses. Can be used on bentgrass, Kentucky bluegrass, fescues (chewing, creeping red, sheeps, tall, and hard) perennial ryegrass, bermudagrass, and zoysiagrass. Do not use on turf mowed at less than 0.5 inches. Repeat applications can be made at 4-week intervals. For a complete list of weeds controlled and additional precautions, consult GameOn herbicide label
Postemergence Control, Black Medic, White, Hop Clover, Buckhorn Plantain, Common Chickweed, Mouseear Chickweed, Henbit, Spurweed (Lawn Burweed), Broadleaf Plantain, Dandelion, False Dandelion, Lespedeza, Prostrate Spurge, Wild Violet					
[triclopyr + cloypyralid], MOA 4 + 4 (3 SL)	Confront, 2-D	0.37 to 0.74 fl oz	1 to 2 pt	0.28 to 0.56 + 0.09 to 0.19	Do not apply to home lawns. May be used on centipedegrass, bermudagrass, zoysiagrass, tall fescue, creeping red fescue, chewing fescue, Kentucky bluegrass, perennial ryegrass. Repeat treatment may be necessary for prostrate spurge and wild violet. Quali-Pro formulation: maintain 0.5 inch height for warm season turf. Do not apply to bermudagrass sod farms. Wait 3 weeks to reseed. Do not use grass clippings for compost or mulch.
[MCPA ester + triclopyr ester + dicamba], MOA 4 + 4 + 4 (3.6 EC)	Cool Power, Lesco Three-Way Ester II, Monterey Spurge Power	0.91 to 1.29 fl oz	2.5 to 3.5 pt	1.125 to 1.575	May be applied to home lawns by a commercial applicator. Not for use on turf grown for resale or other commercial use as sod or seed production. Use on perennial bluegrass, ryegrass, fescue species, bentgrass (excluding greens and tees), bermudagrass, zoysiagrass, and bahiagrass. Do not apply to seedling grasses until well established. Wait 3 to 4 weeks after application to seed.
[MCPA amine + triclopyr amine + dicamba], MOA 4 + 4 + 4 (4.56 L)	Horsepower, Lesco Eliminate	0.73 to 1.1 fl oz	2 to 3 pt	1.14 to 1.71	
[MCPA amine + fluroxypyr ester + triclopyr amine], MOA 4+4+4 (3.41 L)	Battleship III	0.37 to 1.47 fl oz	1 to 4 pt	0.42625 to 1.705	Apply by a commercial applicator to residential, industrial, and institutional lawns, sod farms, parks, cemeteries, athletic fields, roadsides, and golf courses excluding greens and tees. May apply to bentgrass, Kentucky bluegrass, perennial ryegrass, fescue species, bahiagrass, bermudagrass, centipedegrass, and zoysiagrass. Do not spray on warm season turf less than 0.5 inch and do not exceed 3 pints per acre. Generally apply 3 to 4 pints per acre except on fairway bentgrass, which can only tolerate 2 pints per acre. Wait 3 to 4 weeks after application to reseed. Check label for spray adjuvant recommendation.
Postemergence Control, Plantain, Chickweed, Dandelion, Purslane, and Thistle Species, Ground Ivy, Lawn Burweed, Henbit, Corn Speedwell, Spotted Spurge					
carfentrazone-ethyl, MOA 14 (1.9 EW)	Aim EC, QuickSilver T&O	0.0126 to 0.048 fl oz	0.55 to 2.1 fl oz	0.008 to 0.031	May be applied to bahiagrass, bermudagrass, buffalograss, centipedegrass, St. Augustinegrass, zoysiagrass, Kentucky bluegrass, tall fescue, fine fescue, perennial ryegrass, and bentgrass. To expand the weed spectrum and extend control of the weeds listed here and on the label, carfentrazone-ethyl can be tank mixed with the entire range of phenoxy products—amines, esters, and other salts—and is also compatible with dicamba, atrazine, glyphosate, glufosinate, clopyralid, triclopyr, and MSMA. When applied alone, add 0.12 to 0.25% nonionic surfactant.
Postemergence Control, White Clover, Dandelion, Ground Ivy, Spurges, Plantains, Chickweeds, Henbit, Lawn Burweed, Woodsorrels, Dollarweed, Poison Ivy, Poison Oak, Corn Speedwell, Wild Strawberry, Wild Violet, Virginia Pepperweed, Shepherd's Purse					
[carfentrazone + 2,4-D ester + MCPP + dicamba], MOA 14 + 4 + 4 + 4 (2.2 EC)	(2.2 EC): Speed Zone	0.75 to 1.8 fl oz	2 to 5 pt	0.55 to 1.375	May be used on annual and perennial bluegrass, annual and perennial ryegrass, tall and fine fescue, creeping and colonial bentgrass, common and hybrid bermudagrass, and zoysiagrass. For use in ornamental turf, golf courses, lawns, sod farms, cemeteries, and parks. Optimum results when applied when temperatures are between 45 and 75 degrees F but may be applied up to 90 degrees F. Lower rates may be used in cooler weather. Rainfast within 3 hr and may reseed after 2 weeks. May apply 3 to 4 weeks after sodding, sprigging, or plugging. Also may be used on bahiagrass, buffalograss, St. Augustinegrass, centipedegrass, seashore paspalum, and kikuyugrass. May reseed after 1 week.
[carfentrazone + 2,4-D ester + MCPP + dicamba], MOA 14 + 4 + 4 + 4 (0.81 EC)	(0.81 EC): Speed Zone Southern	0.55 to 2.2 fl oz	1.5 to 6 pt	0.1519 to 0.6075	
[carfentrazone + MCPA ester + MCPP + dicamba], MOA 14 + 4 + 4 + 4 (2.91 EC)	Power Zone	0.75 to 2.2 fl oz	2 to 6 pt	0.7275 to 2.1825	Same precautions and turf uses as [carfentrazone + 2,4-D ester + MCPP + dicamba] 2.2 EC except cannot be applied to creeping and colonial bentgrass.
penoxsulam + sulfentrazone + 2,4-D + dicamba	Avenue South	1.0 to 2.2 fl oz	2.7 to 6 pt	0.271 to 0.602	May be used on established Kentucky bluegrass, annual bluegrass, perennial ryegrass, annual ryegrass, tall fescue, common and hybrid bermudagrass, zoysiagrass, centipedegrass, and seashore paspalum. Do not apply more than 2.7 pints per acre on fescue or perennial ryegrass unless turf injury can be tolerated. Do not apply more than 3.5 pints per acre on St. Augustinegrass unless injury, discoloration, stunting, and thinning can be tolerated.

Chemical Weed Control in Lawns and Turf (continued)

Herbicide and Formulation	Brand Names	Amount of Formulation per 1,000 sq ft	Amount of Formulation per Acre	Pounds Active Ingredient per Acre	Precautions and Remarks
Postemergence Control, White Clover, Dandelion, Ground Ivy, Spurges, Plantains, Chickweeds, Henbit, Lawn Burweed, Woodsorrels, Dollarweed, Poison Ivy, Poison Oak, Corn Speedwell, Wild Strawberry, Wild Violet, Virginia Pepperweed, Shepherd's Purse (continued)					
[sulfentrazone + 2,4-D amine + MCPP + dicamba], MOA 14 + 4 + 4 + 4 (2.18 SL)	Surge	0.92 to 1.84 fl oz	2.5 to 5 pt	0.68 to 1.36	Apply 2.5 to 3.25 pints per acre on warm season turf including bermudagrass species, zoysiagrass, bahiagrass, and buffalograss. Apply 3.25 to 4 pints per acre on cool season turf including species of bluegrass, ryegrass, fescue, and bentgrass (excluding greens and tees). 4 to 5 pints per acre needed to control corn speedwell and wild violet. Turf areas include residential, ornamental, institutional, and sod farms. Apply to grass seedlings after second mowing. Apply to sodded, sprigged, or plugged areas 3 to 4 weeks after operations. Treated areas may be reseeded 3 weeks after application.
[triclopyr ester + sulfentrazone + 2,4-D ester + dicamba], MOA 4 + 14 + 4 + 4 (2.51 EC)	Tzone SE	0.75 to 1.5 fl oz	2 to 4 pt	0.628 to 1.26	Apply 2 to 2.25 pints per acre on fully dormant bermudagrass, zoysiagrass, and bahiagrass. Apply 3.25 to 4 pints per acre on annual and perennial bluegrass and ryegrass, and tall, red, and fine fescue. Rainfast within 3 hours. Approved turf areas include residential, ornamental, institutional, noncropland, and sod farms. Apply to grass seedlings after the second or third mowing. Apply to sodded, sprigged, or plugged areas 3 to 4 weeks after operations. Treated areas may be reseeded 3 weeks after application.
Postemergence Control, Chickweed, Clover, Plantain and Dandelion Species, Florida Betony, Dollarweed, Groud Ivy, Lespedeza, and Yellow Woodsorrel					
florasulam, MOA 2 (0.42 SC)	Defendor	0.09 fl oz	4 fl oz	0.013125	Can be used on all established major warm and cool season turfgrass species in residential lawns, golf courses (excluding putting greens), sports fields, sod farms and commercial turf areas. Controls Carolina geranium, species of chickweed, clover and dandelion, vetch, dollarweed and common groundsel. Do not exceed 3 applications or 12 fluid ounces per acre per year. Apply to newly seeded or sprigged turf after third mowing or when tillering and secondary root development has occurred. Wait 4 weeks to reseed. When used alone, add a nonionic surfactant at 0.2% by volume.
Postemergence Control, Chickweed, Clover, Plantain and Dandelion Species, Florida Betony, Dollarweed, Groud Ivy, Lespedeza, and Yellow Woodsorrel (continued)					
penoxsulam, MOA 2 (0.014 G)	(014 G): Harrels Fert. with Penoxsulam	3.4 to 10.3 lb	150 to 450 lb	0.02 to 0.06	May be applied to residential and commercial lawns, golf courses (excluding greens and tees), parks, athletic fields, and sod farms. Use on turf that has been mowed at least 3 times or sprigs that have developed secondary root systems. Apply up to 75 pounds per acre of 0.03 G or 150 pounds per acre of 0.014 G to perennial ryegrass and tall fescue. Apply up to 150 pounds per acre of 0.03 G or 300 pounds per acre of 0.014 G to bentgrass, Kentucky bluegrass, and fine fescue. Apply up to 200 pounds per acre of 0.03 G or 450 pounds per acre of 0.014 G to bermudagrass, centipedegreass, zoysiagrass, and St. Augustinegrass. Do not apply to dormant centipedegrass. Reapply at 4 weeks if needed but do not exceed 300 pounds per acre of 0.03 G or 650 pounds per acre of 0.014 G per season. After treatment, wait 3 to 4 weeks to reseed.

Same statement as above concerning turf uses and reseeding intervals. Bermudagrass and kikuyugrass are the only warm season grasses labeled for use. Apply up to 1 pint per acre on bentgrass, 1.5 pint per acre on bermudagrass and kikuyugrass and 2 pints per acre on tall fescue and perennial ryegrass. Do not apply greater than 2.3 pints per acre per year. Surfactant not required. |
(0.03 G)	(0.03 G): Harrels Fert. with Penoxsulam, Lesco LockUp, Pennington Seed LockUp	1.7 to 4.6 lb	75 to 200 lb		
(0.31 L)	(0.31 L): Sapphire	0.092 to 0.55 fl oz	0.25 to 1.5 pt	0.01 to 0.058	
Carpetweed, Chickweed, Dandelion, Curly Dock, Cutleaf Eveningprimrose, Henbit, Knotweed, Common Mallow, Poison Ivy, and Annual Sowthistle					
pyraflufen ethyl, MOA 14 (0.177 SC)	Octane 2% SC	0.016 to 0.092 fl oz	0.7 to 4 fl oz	0.000938 to 0.0055	Used in established sod farm and ornamental turf by commercial applicators and professional landscapers only. Turf can be newly seeded, sodded, or sprigged as long as it is established and not under stress. Tolerant turfgrasses include bermudagrass, centipedegrass, St. Augustinegrass, zoysiagrass, tall fescue, perennial ryegrass, perennial bluegrass, and creeping bentgrass (not greens or tees). Apply 1 to 4 fluid ounces alone to 3- to 6-inch tall weeds. For larger weeds and broader spectrum control, apply 0.75 to 1.5 fluid ounces and tank mix with 2,4-D, mecoprop, dicamba, chloroprop, MCPA, triclopyr, or fluroxypyr.
Postemergence Control, Bahiagrass, Perennial Ryegrass, Wild Garlic, Spurweed, Henbit, Miscellaneous Other Broadleaf Weeds					
metsulfuron, MOA 2 (60 DF)	Accurate, Amtide MSM 60 DF, Manor, MSM 60 DF, Ortho Weed B Gon Pro St. Augustine, Rometsol	0.003 to 0.02 oz	0.125 to 1 oz	0.005 to 0.038	May be applied to established bermudagrass, zoysiagrass (Meyer or Emerald), St. Augustinegrass, Kentucky bluegrass or fine fescue. Do not apply to turf less than 1 year old. Do not exceed 0.5 ounces per acre on centipedegrass, fine fescue, or Kentucky bluegrass. See label for a complete list of weeds controlled. The addition of 0.25% nonionic surfactant will enhance control. May be used for removal of perennial ryegrass from overseeded warm-season turf species. For bahiagrass control, use 0.25 to 0.75 ounces per acre after spring greenup but before seedhead development. A repeat treatment may be necessary in 4 to 6 weeks.
metsulfuron (Patriot) 60 WDG	Patriot 60 WDG	0.007 to 0.046 oz	0.33 to 2 oz	0.012 to 0.075	Apply to unimproved industrial turf only. Use maximum of 0.5 ounce per acre for fescue and bluegrass and 2 ounces per acre for bermudagrass.
[metsulfuron + rimsulfuron], MOA 2 + 2 (37 WG)	Negate	0.0344352 oz	1.5 oz	0.0346875	See comments under postemergence annual bluegrass control. For bahiagrass control, a repeat treatment may be necessary 4 to 6 weeks after initial application.

Chemical Weed Control in Lawns and Turf (continued)

Herbicide and Formulation	Brand Names	Amount of Formulation per 1,000 sq ft	Amount of Formulation per Acre	Pounds Active Ingredient per Acre	Precautions and Remarks
Postemergence Control, Wild Garlic, Wild Onion					
imazaquin, MOA 2 (70 DG)	Image 70 DG	0.128 to 0.256 oz	0.357 to 0.714 lb	0.25 to 0.5	Use on bermudagrass, centipedegrass, St. Augustinegrass, and zoysiagrass. Do not apply during spring greenup. Temporary yellowing may occur. Add a nonionic surfactant at 2 pints per 100 gallons of spray solution.
2,4-D amine, MOA 4 (4 SL)	Many trade names	2.2 fl oz	3 qt	3	Apply in fall when garlic is young and actively growing. Add a wetting agent to keep spray from bouncing off garlic leaves. Repeat treatment for 2 years. Avoid spray drift which can injure susceptible plants. Use on bluegrass, fescue, bermudagrass, or zoysia. For more susceptible grasses, uses spot treatment below.
		Spot treatment			One tbsp of 1% 2,4-D solution per garlic clump or use pressurized applicator. Apply December to April. Use as spot treatment for widely scattered clumps in small areas. Avoid excessive spraying as turfgrass injury may result.
Postemergence Control of Various Grass and Broadleaf Weeds in Unimproved Turf and Other Noncrop Areas					
glyphosate, MOA 9 (5.5 SL) (5 SL) (4 SL)	Many trade names	0.14 to 1.01 0.12 to 0.87 fl oz 0.75 to 2.94 fl oz	0.375 to 2.75 pt 0.3125 to 2.3751 1 pt to 4 qt	0.26 to 1.89 0.2 to 1.48 0.5 to 4	Check specific labels for correct rates. Apply to dormant or actively growing well established bermudagrass and bahiagrass. Bahiagrass growth will be suppressed if treated after spring greenup and before seedhead formation. Treat winter annual weeds when less than 6 inches tall. Higher rates are needed for more mature plants. Apply in 10 to 40 gallons of water per acre and use an NIS at 2 quarts per 100 gallons of spray solution.
[glyphosate + 2,4-D amine], MOA 9 + 4 (1.2 + 1.9 lb/gal SL)	Campaign	0.55 to 1.47 fl oz	1.5 to 4 pt	0.58 to 1.55	Apply in 15 to 30 gallons of water per acre. May be applied to highly maintained dormant bermudagrass at 2 to 4 pt per acre. In low maintenance bermudagrass, sulfometuron can be added at 0.25 to 1 ounce per acre when dormant or actively growing. Apply 2 to 4 pints per acre on dormant bahiagrass and 1.5 to 2 pints per acre on actively growing bahiagrass. Tank mix with sulfometuron if needed. Check label for sulfometuron rates. Tall fescue applications can be made in the spring or summer at 2 to 3 pints per acre with or without sulfometuron. Spray tall fescue at 4 to 6 inches tall and before seedhead emergence to minimize injury.
Postemergence Control of Various Grass and Broadleaf Weeds in Unimproved Turf and Other Noncrop Areas (continued)					
sulfosulfuron, MOA 2 (75 WG)	Certainty, Outrider	0.017 to 0.046 oz	0.75 to 2 oz	0.035 to 0.094	May be used in well-established dormant and actively growing bermudagrass and bahiagrass. Wait 30 days to re-treat if needed; do not exceed 2.66 ounces per acre per year. If treating weeds postemergence, use an NIS at 2 quarts per 100 gallons spray solution unless tank mixed with glyphosate. Sulfosulfuron can be tank mixed with [glyphosate + 2,4-D amine], metsulfuron, sulfometuron, and chlorsulfuron, but check label for proper turf species and timing. Expect temporary injury or discoloration with tank mix partners. For well-established tall fescue, do not exceed 1 ounce per acre per year, and do not tank mix. Effective on johnsongrass.
[thiencarbazone-methyl + iodosulfuron-methyl + foramsulfuron], MOA 14 + 2 + 2 (36.4 WDG)	Derigo	0.069 to 0.138 oz	3 to 6 oz	0.068 to 0.137	May be applied to unimproved bermudagrass, zoysiagrass, centipedegrass and bare ground sites on private, public and military land for control of many annual and perennial broadleaf and grass weeds. Check label for complete weed listing, rate needed and recommended adjuvant. Repeat application in 4 to 6 weeks if weed regrowth occurs not to exceed 6 ounces product per year. Spot treatment (spray-to-wet) rate is 3 to 6 ounces product per 25- to 100-gallon solution. Nonionic surfactant is generally recommended at 0.25 to 0.5%. Use 0.5 to 1% methylated seed oil for difficult to control broadleaf weeds and perennial grasses. Increased control may be achieved with 1.5 to 3 pounds per acre ammonium sulfate in high humidity climates or 1.5 to 2 quarts per acre urea ammonium nitrate in low humidity climates. Do not use an organosilicone surfactant.
Postemergence Control in Dormant Warm Season Turf Annual Bluegrass, Various Other Winter Annual Weeds					
diquat, MOA 22 (2 SL)	Diquash, Diquat., Diquat SPC, Harvester, Priceto Diquat 2 L, Reward LS, Tribune	0.4 to 0.75 fl oz	1 to 2 pt	0.25 to 0.5	Apply in 20 to 100 gallons spray mix as a broadcast application. Add 1 to 2 pints of a nonionic surfactant per 100 gallons of solution. Bermudagrass must be dormant. More than one application may be needed.
flumioxazin, MOA 14 (51 WG)	Sureguard	0.1837 to 0.2755 oz	8 to 12 oz	0.255 to 0.3825	Use on completely dormant bermudagrass turf including residential and commercial lawns, golf courses (excluding greens), sod farms, roadsides, athletic fields, parks and schools. Add 0.25% by volume nonionic surfactant for postemergence applications. Provides preemergence control of annual grasses such as crabgrass, goosegrass, foxtail species, barnyardgrass and annual bluegrass. Does control annual bluegrass postemergence along with many common winter annual broadleaf weeds such as chickweed species, henbit, Carolina geranium and hairy bittercress. Allow a 15 feet buffer zone when applying upslope from bentgrass greens or bermudagrass greens overseeded with *Poa trivialis*. To limit potential lateral movement, do not apply to saturated soil.
glyphosate - Roundup, MOA 9 (4 SL) (5 SL) (5.5 SL)	Many trade names	0.37 fl oz 0.29 fl oz 0.27 fl oz	1 pt 0.8 pt 0.73 pt	0.5	Check specific labels for correct rates. Apply in 5 to 40 gallons water per acre with 0.5% by volume of a nonionic surfactant. Application to actively growing annual bluegrass must be made before initiation of bermudagrass greenup in the spring.

Chemical Weed Control in Lawns and Turf (continued)

Herbicide and Formulation	Brand Names	Amount of Formulation per 1,000 sq ft	Amount of Formulation per Acre	Pounds Active Ingredient per Acre	Precautions and Remarks
Postemergence Control in Dormant Warm Season Turf Annual Bluegrass, Various Other Winter Annual Weeds (continued)					
glyphosate - Touchdown, MOA 9 (3 LC)	(3 SL): Touchdown Pro	0.18 to 1.47 fl oz	0.5 to 4 pt	0.1875 to 1.5	Apply to dormant bermudagrass and bahiagrass before spring greenup. Apply in 10 to 40 gallons water per acre. Will control winter annual weeds up to 6 inches tall and 4- to 6-leaf tall fescue. Use a 75% active ingredient nonionic surfactant at 0.25% by volume or dry ammonium sulfate at 0.5% by weight.
(4.17 LC)	(4.17 SL): Touchdown Total	0.13 to 1.06 fl oz	0.36 to 2.88 pt		
(5 LC)	(5 SL): Touchdown HiTech	0.11 to 0.88 fl oz	0.3 to 2.4 pt		
[glyphosate + diquat], MOA 9 + 22 (76 WG) (4.21 SL)	Razor Burn, Roundup QuikPRO	0.11 to 0.37 oz 0.18 to 0.62 fl oz	5 to 16 oz 8 to 27 fl oz	0.24 to 0.76 0.26 to 0.89	Apply to dormant bermudagrass and bahiagrass not grown for research, sale, or other commercial uses, such as sod, seed production. Apply in 10 to 80 gallons water per acre. Rates greater than 9 ounces per acre of 76 WG product or 15 fluid ounces per acre of 4.21 SL product may cause injury or delay greenup in highly maintained areas. Controls tall fescue.
metribuzin, MOA 5 (75 WDF)	(75 DF): Dimetric DF, Glory, Metribuzin 75, Tricor DF	0.25 oz	0.67 lb	0.5	For application by commercial applicators to dormant bermudagrass turf. Broadcast spray before greenup of turf. Do not apply to greens, tees, or aprons. Controls common chickweed, corn speedwell, henbit, parsley-piert, and spurweed.
Suppression/Control, Bermudagrass					
fenoxaprop, MOA 1 (0.57 EC)	Acclaim Extra	0.46 fl oz	1.25 pt	0.089	Use on Kentucky bluegrass, perennial ryegrass, fine and tall fescue, and zoysiagrass. Apply June 1, July 1, Aug. 1, Sept. 1, repeat for 2 years. Can be tankmixed with 1 pt per acre triclopyr following the same schedule as above. Apply June 1 and Aug. 1 for 2 years if tank mixed with 1 quart per acre triclopyr. Zoysia may show discoloration but should recover in 10 to 14 days following tankmix applications.
fluazifop, MOA 1 (2 EC)	Fusilade II	0.05 to 0.14 fl oz	2 to 6 fl oz	0.03 to 0.09	Use on tall fescue or zoysia. For fescue, apply 5 to 6 ounces per acre during warm weather in early spring when bermudagrass is breaking dormancy; repeat in fall when bermudagrass is preparing for dormancy. For zoysia, apply 4 ounces per acre on June 1, Aug. 1; repeat for 2 years. Can tank-mix with 1 quart per acre triclopyr following schedule above. Zoysia or tall fescue may show slight discoloration but should recover in 10 to 14 days. Add a nonionic surfactant at 0.25% v/v. Apply in a minimum of 30 gallons of water per acre.
Suppression/Control, Bermudagrass (continued)					
siduron, MOA 7 (50 WP)	Tupersan	0.5 to 1 lb	21.78 to 43.56 lb	10.88 to 21.78	Apply as 8- to 12-inch band treatment with a single nozzle sprayer along putting green perimeter to suppress bermudagrass stolon encroachment. Initiate in March or April, and continue subsequent applications at 4- to 5-week intervals.
triclopyr, MOA 4 (4 EC)	Monterey Turflon Ester, Remedy Ultra, Turflon Ester Ultra	0.73 fl oz	1 qt	1.0	Use on perennial bluegrass, perennial ryegrass, tall fescue or ornamental turf including sod farms and golf courses. Do not apply to zoysia unless injury can be tolerated. Apply June 1, July 1, Aug. 1, Sept. 1, repeat for 2 years. Can be tank-mixed with fenoxaprop or fluazifop at rates, timings listed above. New low-odor formulation uses methylated seed oil solvents instead of petroleum distillates.
Postemergence Control Bermudagrass					
clethodim, MOA 1 (0.97 EC)	(0.97 EC): Envoy Plus, TapOut	0.4 to 0.8 oz	17 to 34 fl oz	0.125 to 0.25	For use on sod farms only. Do not apply to centipedegrass being grown for seed. Do not apply until 3 weeks after full greenup of centipedegrass in spring. Do not mow for 1 week before and after application. The addition of a nonionic surfactant at 0.25 % solution (1 pint per 50 gallons water) or a crop oil concentrate at 1% solution (2 quart per 50 gallons water) is necessary for control. A repeat application usually 3 to 4 weeks after the first application will be required for bermudagrass control. Use higher rates for more established bermudagrass. Do not apply more than 68 ounces of clethodim per acre per year. Some discoloration of centipedegrass will occur at higher rates.
Preplant Control or Lawn Renovation — Emerged Annual and Perennial Grass and Broadleaf Weeds					
glyphosate - Roundup, MOA 9 (4 SL) (5 SL) (5.5 SL)	Many trade names	0.75 to 3 oz 0.54 to 2.17 fl oz 0.54 to 2.14 fl oz	1 to 4 qt 0.8 to 3.2 qt 0.73 to 2.91 qt	1 to 4	Where existing vegetation is growing in a field or unmowed situation, apply to actively growing weeds at the stages according to label. Where existing vegetation is growing under mowed turfgrass management, apply after omitting at least one regular mowing to allow sufficient growth for good interception of the spray. Tillage or renovation techniques such as vertical mowing, coring, or slicing should be delayed for 7 days after application. Desirable turfgrass may be established following treatment.
glyphosate - Touchdown, MOA 9 (3 LC) (4.17 LC) (5 LC)	(3 SL): Touchdown Pro (4.17 SL): Touchdown Total (5 SL): Touchdown HiTech	0.18 to 1.47 fl oz 0.13 to 1.06 fl oz 0.11 to 0.88 fl oz	0.5 to 4 pt 0.36 to 2.88 pt 0.3 to 2.4 pt	0.1875 to 1.5	Same remarks as glyphosate, above. In addition, use a 75% active ingredient nonionic surfactant at 0.25% by volume or dry ammonium sulfate at 0.5% by weight.

Chemical Weed Control in Lawns and Turf (continued)

Herbicide and Formulation	Brand Names	Amount of Formulation per 1,000 sq ft	Amount of Formulation per Acre	Pounds Active Ingredient per Acre	Precautions and Remarks
Preplant Control or Lawn Renovation — Emerged Annual and Perennial Grass and Broadleaf Weeds (continued)					
[glyphosate + diquat], MOA 9 + 22 (76 WG) (4.21 SL)	Razor Burn, Roundup QuikPRO	1.65 to 4.5 oz 2.75 to 5.5 fl oz	4.5 to 12.25 lb 3.75 to 7.5 qt	3.4 to 9.3 3.95 to 7.89	Generally use the 75 WG product at 4.5 pounds per acre on annuals, 9 pounds per acre on perennials, and 12.25 pounds per acre on dusty or stressed plants, dense stands, or difficult-to-control perennials. Generally use the 4.21 SL product at 3.75 quarts per acre on annuals and 7.5 quarts per acre on perennials. Do not use on turf grown for research, for sale, or for commercial uses, such as sod or seed production. Do not use if renovating bermudagrass or kikuyugrass sods. Delay tillage for 7 days after application.
[indaziflam + diquat dibromide + glyphosate], MOA 21 + 22 + 9 (1.958 SL)	Specticle Total	1 pt	5.44 gal	10.66	For nonselective preemergence and postemergence control in noncrop areas. Reapply 4 months after initial application if needed not to exceed 1 quart per 1000 square feet per year. Apply 1 pint in 1 gallon of water to cover 1000 square feet. Do not seed for 12 months after application.
Dazomet 99G	Basamid 99G	6 lb	262 lb		Restricted use pesticide. Must be US EPA certified applicator. A medically fitted respirator is needed for application. Apply 262 pounds product per acre and follow label recommendations for water and/or mechanical incorporation. Conrol is improved significantly by applying glyphosate (3 lb/A) + fluazifop (0.4 lb/A) 7 days prior to dazomet application.
Trimming and Edging and Control of Emerged Weeds					
diquat, MOA 22 (2 SL)	Diquash, Diquat., Diquat SPC, Harvester, Priceto Diquat 2 L, Reward LS, Tribune	0.4 to 0.75 fl oz	1 to 2 pt	0.25 to 0.5	Add nonionic surfactant at 0.25 ounce per gallon of water. Water volumes above 15 gal per acre should be used. For spot sprays, use 0.3 to 0.75 fluid ounce per gallon.
glufosinate, MOA 10 (1 SL)	Finale	2.2 to 4.4 fl oz	3 to 6 qt	0.75 to 1.5	Rate depends on weed to be controlled and stage of growth. Consult label. For spot or directed spray use 1.5 to 4 fluid ounces per gallon of water.
glyphosate + diquat, MOA 9 + 22 (76 WG) (4.21 SL)	Razor Burn, Roundup QuikPRO	1.65 to 4.5 oz 2.75 to 5.5 fl oz	4.5 to 12.25 lb 3.75 to 7.5 qt	3.4 to 9.3 3.95 to 7.89	May be used in general noncrop areas. Do not use on plants grown for sale or other commercial uses, such as seed production. See rate comments in lawn renovation section. For spray to wet treatments, apply the 76 WG product at 1.2 ounces per gal of water for annuals and 1.5 ounces per gal of water for perennials. Apply the 4.21 SL product at 2 fluid ounces per gallon of water for annuals and 2.5 fluid ounces per gal water for perennials. For directed spot treatment of perennials using hand-held low volume equipment, apply 4 to 8 ounces per gallon of water.

Herbicide Modes of Action for Hay Crops, Pastures, Lawns and Turf

Brands listed were registered for sale and use in North Carolina in 2021 according to www.kellysolutions.com/nc/searchbychem.asp. Active ingredients enclosed within parentheses are prepackaged herbicides.

Herbicide Modes of Action for Hay Crops, Pastures, Lawns and Turf

Active Ingredient(s)	Brand Names	Chemical Family	Mode of Action[1]
(2, 4-D + mecoprop + dichlorprop)	Spoiler, Triamine	phenoxycarboxylic acid + phenoxyalkanoic acid + chlorinated phenoxy	4 + 4 + 4
2,4-D amine	(4 SL): various trade names	phenoxy-carboxylic acid	4
(2,4-D + aminopyralid)	GrazonNext HL	phenoxy-carboxylic acid + pyridinecarboxylic acid	4 + 4
(2,4-D ester + carfentrazone-ethyl)	Rage D-Tech	phenoxy-carboxylic acid + triazolinone	4 + 14
(2,4-D + clopyralid + dicamba)	Millennium Ultra 2	phenoxycarboxylic acid + pyridinecarboxylic acid + benzoic acid	4 + 4 + 4
(2,4-D amine + cloypyralid)	Curtail	phenoxy-carboxylic acid + pyridinecarboxylic acid	4 + 4
(2,4-D + dicamba)	Brash, Range Star, Rifle-D, Weedmaster	phenoxy-carboxylic acid + benzoic acid	4 + 4
(2,4-D + fluroxypyr + dicamba)	E-2, Escalade 2	phenoxy + pyridinecarboxylic acid + benzoic acid	4 + 4 + 4
(2,4-D + glyphosate)	Campaign	phenoxy-carboxylic acid + glycine	4 + 9
(2,4-D + mecoprop + dicamba)	3-D, Three-Way Selective Herbicide, Triplet SF, various Trimec formulations	phenoxycarboxylic acid + phenoxyalkanoic acid + benzoic acid	4 + 4 + 4
(2,4-D + picloram)	Grazon P+D, Trooper P+D	phenoxy-carboxylic acid + pyridinecarboxylic acid	4 + 4
(2,4-D + triclopyr)	Candor, Crossbow	phenoxy-carboxylic acid + pyridinecarboxylic acid	4 + 4
2,4-DB	(1.75 EC): 2,4-DB 175 (2 EC): 2,4-DB 200, Butyrac 200	phenoxy-carboxylic acid	4
amicarbazone	Xonerate	triazolinone	5
aminopyralid	Milestone	pyradinecarboxylic acid	4
(aminopyralid + metsulfuron methyl)	Chaparral	pyradinecarboxylic acid + sulfonylurea	4 + 2
asulam	Asulox	carbamate	18
atrazine	(4 L): AAtrex 4 L, Atrazine 4 L (90 DF, 90 WDG): Aatrex Nine-O, Atrazine 90 WDG, Atrazine 90 DF	triazine	5
benefin	Lebanon Balan 2.5 G, The Andersons Crabgrass Preventer with 2.5 Balan	dinitroaniline	3
(benefin + trifluralin)	Fertilizer with Team Pro 0.86%	dinitroaniline + dinitroaniline	3 + 3
bensulide	(4 EC): Bensumec 4 LF (8.5 G): Weedgrass Preventer (12.5 G): Pre-San Granular	organophosphorus	8
(bensulide + oxadiazon)	Goosegrass / Crabgrass Control	organophosphorus + oxadiazole	8 + 14
bentazon	Basagran Sedge Control, Basagran T/O, Bentazon 4, Lescogran	benzothiadiazole	6
bispyribac-sodium	Velocity	pyrimidinyloxybenzoic acid	2
bromoxynil	(2 EC): Broclean, Maestro 2 EC (4 EC): Buctril 4 EC	nitrile	6
carfentrazone-ethyl	Aim EC, QuickSilver T&O	triazinone	14
(carfentrazone + 2,4-D ester + mecoprop + dicamba)	(2.2 EC): Speed Zone (0.81 EC): Speed Zone Southern	triazinone + phenoxycarboxylic acid + phenoxyalkanoic acid + benzoic acid	14 + 4 + 4 + 4
(carfentrazone + MCPA + mecoprop + dicamba)	Power Zone	triazinone + phenoxy + phenoxyalkanoic acid + benzoic acid	14 + 4 + 4 + 4
(carfentrazone + quinclorac)	SquareOne	triazinone + quinoline carboxylic acid	14 + (27 + 4)
chlorsulfuron	Alligare Chlorsulfuron 75, Telar XP	sulfonylurea	2

Herbicide Modes of Action for Hay Crops, Pastures, Lawns and Turf (continued)

Active Ingredient(s)	Brand Names	Chemical Family	Mode of Action[1]
clethodim	(2 EC): Arrow 2 EC, Avatar S2, Cleanse 2 EC, Clethodim 2 LC, Clethodim 2 E, Dakota, Section 2 EC, Shadow, Tide Clethodim 2 EC, Select 2 EC, Volunteer, Willowood Clethodim 2 EC (0.97 EC): Envoy Plus, TapOut	cyclohexanedione	1
cloypyralid	Lontrel T&O	pyridinecarboxylic acid	4
dicamba / diglycolamine	Banvel, Clarity, Clash, Detonate, Rifle, Strut, Topeka, Vanquish	benzoic acid	4
diclofop-methyl	Illoxan	aryloxyphenoxy propionate	1
(diflufenzopyr-sodium + dicamba)	Overdrive	semicarbazone + benzoic acid	19 + 4
dimethenamid	Tower	chloroacetamide	15
(dimethenamid + pendimethalin)	Freehand	chloroacetamide + dinitroaniline	15 + 3
diquat	Diquash, Diquat., Diquat SPC, Harvester, Priceto Diquat 2 L, Reward LS, Tribune	bipyridylium	22
diuron	Direx 4 L, Diuron 4 L	phenylurea	7
dithiopyr	(2 EW, 2 L): Armortech CGC 2 L, Dimension 2 EW, Dithiopyr 2 L (40 WP): Armortech CGC 40, Dimension Ultra, Dithiopyr 40 WSP	pyridine	4
EPTC	Eptam 7-E	thiocarbamate	8
ethofumesate	(1.5 EC): Progress (4 EC): Phoenix Thrasher, PoaConstrictor, Progress SC	benzofuranes	8
fenoxaprop	Acclaim Extra	aryloxyphenoxy propionate	1
flazasulfuron	Katana	sulfonylurea	2
florasulam	Defendor	triazolopyrimidine	2
fluazifop	Fusilade II	aryloxyphenoxy propionate	1
flumioxazin	Sureguard	N-phenylphthalimide	14
foramsulfuron	Revolver	sulfonylurea	2
glufosinate	Finale	organophosphorus	10
glyphosate – Roundup formulations	(4 SL, 5 SL, 5.4 SL, 5.5 SL): various trade names	glycine	9
glyphosate – Touchdown formulations	(3 SL): Touchdown Pro (4.17 SL): Touchdown Total (5 SL): Touchdown HiTech		
(glyphosate + diquat)	Razor Burn, Roundup QuikPRO	glycine + bipyridylium	9 + 22
halosulfuron	HiYield Nutsedge Control, Nutgrass Killer II, Profine 75, Sandea	sulfonylurea	2
imazapic	Imazapic 2 SL, Impose, Panoramic, Plateau	imadazolinone	2
imazaquin	Image 70 DG	imidazolinone	2
imazethapyr	Pursuit, Slay, Thunder	Imidazolinone	2
imazosulfuron	Celero	sulfonylurea	2
indaziflam	(20 WSP): Specticle 20 WSP (0.0224 G): Specticle G (0.622F): Specticle Flo	benzamide	21
(indaziflam + diquat bromide + glyphosate)	Specticle Total	benzamide + bipyridylium + glycine	21 + 22 + 9
isoxaben	Gallery, Isoxaben	benzamide	21
(MCPA + mecoprop + dicamba)	Ortho Weed B Gon Pro Southern, Tri-Power	mcpa + phenoxyalkanoic acid + benzoic acid	4 + 4 + 4
(MCPA amine + fluroxypyr ester + dicamba)	Change Up	phenoxy + pyridinecarboxylic acid + benzoic acid	4 + 4 + 4
(MCPA amine + fluroxypyr ester + triclopyr amine)	Battleship III	phenoxy + pyridinecarboxylic acid + pyridinecarboxylic acid	4 + 4 + 4
(MCPA amine + triclopyr amine + dicamba)	Horsepower, Lesco Eliminate	phenoxy + pyridinecarboxylic acid + benzoic acid	4 + 4 + 4

Herbicide Modes of Action for Hay Crops, Pastures, Lawns and Turf (continued)

Active Ingredient(s)	Brand Names	Chemical Family	Mode of Action[1]
(MCPA ester + triclopyr ester + dicamba)	Cool Power, Lesco Three-Way Ester II, Monterey Spurge Power	phenoxy + pyridinecarboxylic acid + benzoic acid	4 + 4 + 4
mecoprop	MCPP-p4 Amine, Mecomec 2.5, Mecomec 4	phenoxyalkanoic acid	4
mesotrione	Tenacity	benzoylcyclohexanedione	27
metolachlor	Pennant Magnum	chloroacetamide	15
metribuzin	(75 DF): Dimetric DF, Glory, Metribuzin 75, Tricor DF 4 L, 4 F): Glory 4L, Tricor 4 F	triazinone	5
metsulfuron methyl	Accurate, Amtide MSM 60 DF, Amitide MSM Turf, Manor, MSM 60 DF, MSM Turf, Ortho Weed B Gon Pro St. Augustine, Plotter, Purestand, Rometsol	sulfonylurea	2
(metsulfuron + 2,4-D + dicamba)	Cimarron Max	sulfonylurea + phenoxy-carboxylic acid + benzoic acid	2 + 4 + 4
(metsulfuron + chlorsulfuron)	Chisom, Cimarron Plus	sulfonylurea + sulfonylurea	2 + 2
(metsulfuron methyl + rimsulfuron)	Negate	sulfonylurea + sulfonylurea	2 + 2
monosodium methylarsonate	(6 SL): Target 6 Plus, Drexel MSMA 6 Plus (6.6 SL): Target 6.6 Plus, Drexel MSMA 6.6 Plus	organic arsenical	17
napropamide	Devrinol 50 DF	acetamide	15
(nicosulfuron + metsulfuron methyl)	Pastora	sulfonylurea + sulfonylurea	2 + 2
oryzalin	(4 AS, 4 L): Monterey Weed Impede 4 AS, Oryzalin 4 AS, Phoenix Harrier 4 L, Surflan AS (85WDG): Surflan WDG	dinitroaniline	3
oxadiazon	(2 G): Oxadiazon 2 G, Ronstar G (50 WP): Oxadiazon 50 WSB, Ronstar 50 WSB (3.17 SC): Oxadiazon SC, Phoenix Starfighter L, Ronstar Flo	oxadiazole	14
(oxadiazon + prodiamine)	Pro-mate Ronstar + Barricade 1.2 G, Regalstar II	oxadiazole + dinitroaniline	14 + 3
paraquat	(2 SL): Cyclone SL 2.0, Gramoxone Inteon, Gramoxone SL (3 SL): Bonedry, Firestorm, Helmquat 3 SL, Parazone 3 SL	bipyridylium	22
pendimethalin	(3.8 CS): HydroCap, Pendulum AquaCap, Pre-M AquaCap, Prowl H2O, Satellite HydroCap (3.3 EC): Drexel Pin-Dee 3.3 T&O (2 G): Pendulum 2 G (0.86 G): fertilizers – Pendimethalin, Pre-M, Propendi, Pro-mate Pendi (1.29 G): Step 1 Crabgrass Preventer, Turf Builder with Halts	dinitroaniline	3
penoxsulam	(0.31 L): Sapphire (0.03 G): Harrels Fert. with Penoxsulam, lesco LockUp, Pennington Seed LockUp (014 G): Harrels Fert. with Penoxsulam	triazolopyrimidine	2
(picloram + fluroxypyr)	Surmount	pyradinecarboxylic acid + pyradinyloxyacetic acid	4 + 4
prodiamine	(65 WDG): Armortech Kade, Barricade, Cavalcade, Halts Pro, Phoenix Knighthawk, ProClipse, Prodiamine, Quali-Pro Prodiamine, RegalKade, Resolute, Stonewall (4 FL): Barricade, Evade, Resolute (0.5 G): RegalKade, Signature Crabgrass Preventer, Turf Pride	dinitroaniline	3

Herbicide Modes of Action for Hay Crops, Pastures, Lawns and Turf (continued)

Active Ingredient(s)	Brand Names	Chemical Family	Mode of Action[1]
pronamide	(50 WP): Kerb 50-W (3.3 SC): Kerb SC T&O, Willowood Pronamide 3.3SC	benzamide	3
pyraflufen ethyl	Octane 2% SC	phenylpyrazole	14
quinclorac	(75 DF): Armortech Quinclorac Pro, Quinclorac, Quinclorac SPC, Quinstar (1.5 SL): Drive XLR8, Quinclorac 1.5 L	quinoline carboxylic acid	(27 + 4)
(quinclorac + mecoprop + dicamba)	Onetime	quinoline carboxylic acid + phenoxyalkanoic acid + benzoic acid	(27 + 4) + 4 + 4
(quinclorac + sulfentrazone + 2,4-D amine + dicamba)	Q4 Plus	quinoline carboxylic acid + triazinone + phenoxy-carboxylic acid + benzoic acid	(27 + 4) + 14 + 4 + 4
rimsulfuron	Rimsulfuron	sulfonylurea	2
sethoxydim	(1.5 EC): Poast (1 EC): Poast Plus, Segment, Sethoxydim SPC	cyclohexanedione	1
siduron	Tupersan	phenylurea	7
simazine	(4 L): Princep Liquid, Simazine, Sim-Trol (90 DF, 90 WDG): Simazine, Sim-Trol, Princep Caliber 90	triazine	5
sulfentrazone	Dismiss Turf	triazinone	14
(sulfentrazone + 2,4-D + mecoprop + dicamba)	Surge	triazinone + phenoxycarboxylic acid + phenoxyalkanoic acid + benzoic acid	14 + 4 + 4 + 4
(sulfentrazone + imazethapyr)	Dismiss South	triazinone + imidazolinone	14 + 2
(sulfentrazone + metsulfuron-methyl)	Blindside	triazinone + sulfonylurea	14 + 2
(sulfentrazone + prodiamine)	Echelon	triazinone + dinitroaniline	14 + 3
(sulfentrazone + quinclorac)	Solitare	triazinone + quinoline carboxylic acid	14 + (27 + 4)
sulfosulfuron	Certainty, Outrider	sulfonylurea	2
tebuthiuron	(20 P): Spike 20 P, Alligare 20 P (80 DF, 80 WG): Spike 80 DF, Alligare 80 WG	thiadiazolyurea	7
terbacil	Sinbar 80 WDG	uracil	5
(thiencarbazone + foramsulfuron + halosulfuron)	Tribute Total	triazolinone + sulfonylurea + sulfonylurea	14 + 2 + 2
(thiencarbazone + iodosulfuron + dicamba)	Celsius WG	triazolinone + sulfonylurea + benzoic acid	14 + 2 + 4
(thiencarbazone + iodosulfuron + foramsulfuron)	Derigo	triazolinone + sulfonylurea + sulfonylurea	14 + 2 + 2
topramezone	Pylex	benzoylpyrazole	27
triclopyr	Monterey Turflon Ester, Remedy Ultra, Turflon Ester Ultra	pyradinecarboxylic acid	4
(triclopyr + cloypyralid)	Confront, 2-D	pyradinecarboxylic acid + pyradinecarboxylic acid	4 + 4
(triclopyr + fluroxypyr)	PastureGard HL	pyradinecarboxylic acid + pyradinyloxyacetic acid	4 + 4
(triclopyr + sulfentrazone + 2,4-D + dicamba)	Tzone SE	pyridinecarboxylic acid + triazinone + phenoxy-carboxylic acid + benzoic acid	4 + 14 + 4 + 4
trifloxysulfuron	Monument	sulfonylurea	2
trifluralin	(10 G): Treflan TR-10, Trifluralin 10 G (4 EC): Treflan HFP, Trifluralin 4 EC	dinitroaniline	3

Tolerance of Established Cool-Season Turfgrasses to Preemergence Herbicides for Control of Annual Weedy Grasses

KEY: T = tolerant when used properly according to the label; M = marginally tolerant, may cause injury or thinning of the turf; NR = not registered for use on this turfgrass. Apply only to established grasses.

Herbicide	Kentucky Bluegrass	Tall Fescue	Fine Fescue	Perennial Ryegrass	Bentgrass Golf Greens
Benefin*	T	T	M	T	NR
Benefin + trifluralin	T	T	M	T	NR
Bensulide*	T	T	T	T	T
Bensulide + oxadiazon	T	T	NR	T	T
DCPA*	T	T	M	T	NR
Dimethenamid	T	T	T	T	NR
Dithiopyr**	T	T	T	T	T
Indaziflam	NR	NR	NR	NR	NR
Metolachlor	NR	NR	NR	NR	NR
Napropamide	NR	T	T	NR	NR
Oryzalin	NR	T	NR	NR	NR
Oxadiazon*	T	T	NR	T	NR
Pendimethalin	T	T	T	T	NR
Prodiamine	T	T	T	T	NR
Siduron***	T	T	T	T	M
Simazine and atrazine	NR	NR	NR	NR	NR

*Only benefin, bensulide, DCPA, and oxadiazon may be applied in the spring to grasses seeded the previous fall.

**Do not use dithiopyr on Chewings fescue, colonial bentgrass, or unamended golf greens.

*** Siduron may be applied when seeding tolerant grasses.

Tolerance of Established Warm-Season Turfgrasses to Preemergence Herbicides for Control of Annual Weedy Grasses

KEY: T = tolerant when used properly according to the label; NR = not registered for use on this turfgrass.

Herbicide	Bahiagrass	Bermudagrass	Bermudagrass Putting Greens	Centipedegrass	St. Augustinegrass	Zoysiagrass
Benefin	T	T	NR	T	T	T
Benefin + trifluralin	T	T	NR	T	T	T
Bensulide	T	T	T	T	T	T
Bensulide + oxadiazon	NR	T	T	NR	NR	T
DCPA	T	T	NR	T	T	T
Dimethenamid	T	T	NR	T	T	T
Dithiopyr*	T	T	T	T	T	T
Indaziflam	T	T	NR	T	T	T
Metolachlor	T	T	NR	T	T	T
Napropamide	T	T	NR	T	T	NR
Oryzalin	T	T	NR	T	T	T
Oxadiazon	NR	T	NR	NR	T	T
Pendimethalin	T	T	T	T	T	T
Prodiamine	T	T	NR	T	T	T
Siduron	NR	NR	NR	NR	NR	T
Simazine and atrazine	NR	T	NR	T	T	T

*Do not use dithiopyr on unamended golf greens.

Tolerance of Turfgrasses to Postemergence Herbicides for Broadleaf Weed Control

Herbicide	Cool-Season					Warm-Season				
	Bentgrass	Kentucky Bluegrass	Tall Fescue	Fine Fescue	Perennial Ryegrass	Bahiagrass	Bermudagrass	Centipedegrass	St. Augustinegrass	Zoysiagrass
2,4-D	S-I	T	T	T	T	S-I	T	S-I	S-I	T
Mecoprop	T	T	T	T	T	T	T	T	S-I	T
Dicamba	S-I	T	T	T	T	T	T	I	S-I	T
2,4-D + Mecoprop	I	T	T	T	T	T	T	S-I	S-I	S-I
2,4-D + Dichlorprop	S-I	T	T	T	T	I	T	S-I	S-I	S-I
2,4-D + Mecoprop + Dicamba	I	T	T	T	T	I	I	S-I	S-I	T
2,4-D + Mecoprop + Dichlorprop	I	T	T	I	T	T	T	I	I	T
Mecoprop + Dichlorprop	I	T	T	T	T	T	T	I	I	T
Bentazon + Atrazine	S*	S	S	S	S	S	T	T	T	T
Florasulam	T*	T	T	T	T	T	T	T	T	T
Penoxsulam	T*	T	T	T	T	T	T	T	T	T
Triclopyr	S	T	T	S	T	S	S	S	S	I
Clopyralid	T	T	T	T	T	T	T	T	T	T
Fluroxypyr	T	T	T	T	T	T	T	T	T	T
2,4-D + Triclopyr	S-I	T	T	I	T	S	S	S	S	S
MCPA + Fluroxypyr + Dicamba	T*	T	T	T	T	T	T	T	I	T
MCPA + Fluroxypyr + Triclopyr	I	T	T	T	T	T	T	T	S	T
MCPA + Triclopyr + Clopyralid	T	T	T	T	T	T	T	T	S	S
MCPA + Triclopyr + Dicamba	T*	T	T	T	T	T	T	T	S	S
Triclopyr + Clopyralid	S	T	T	S	T	S	T	T	S	T
2,4-D + Clopyralid + Dicamba	T*	T	T	T	T	T	T	S	S	T
2,4-D + Fluroxypyr + Dicamba	T*	T	T	T	T	T	T	S	S	T
Carfentrazone	T	T	T	T	T	T	T	T	T	T
Sulfentrazone	T	T	T	T	T	T	T	T	S	T
Carfentrazone + 2,4-D + Mecoprop + Dicamba	T	T	T	T	T	I	T	I	I	T
Sulfentrazone + 2,4-D + Mecoprop + Dicamba	T*	T	T	T	T	T	T	S	S	T
Thiencarbazone + Iodosulfuron + Dicamba	S	S	S	S	S	S	T	T	T	T

KEY: I = intermediate tolerance, use with caution, use at reduced label rates or minimum label rates; S = sensitive, DO NOT USE THIS HERBICIDE; T = tolerant. Since tolerance frequently depends upon rate used and formulation selected, CAREFULLY CHECK THE LABEL.

* Do not use on golf course greens or tees.

Tolerance of Turfgrasses to Postemergence Herbicides for Control of Grass or Broadleaf Weeds

Herbicide	Cool-Season					Warm-Season				
	Bentgrass	Kentucky Bluegrass	Tall Fescue	Fine Fescue	Perennial Ryegrass	Bahiagrass	Bermudagrass	Centipedegrass	St. Augustinegrass	Zoysiagrass
CMA, DSMA, MSMA	I	I	I	I	T	S	T	S	S	I
Asulam	S	S	S	S	S	S	T*	S	T	S
Atrazine	S	S	S	S	S	S	T	T	T	T
Bentazon	T	T	T	T	I	T	T	T	T	T
Bispyribac-sodium	I	S	S	S	I	S	S	S	S	S
Chlorsulfuron	T	T	S	T	S	I	T	I	I	I
Diclofop	S	S	S	S	S	S	T	S	S	S
Dithiopyr	T	T	T	T	T	T	T	T	T	T
Ethofumesate	I	T	T	S	T	S	D	S	T	S
Fenoxaprop	S-I	T	T	T	T	S	S	S	S	T
Flazasulfuron	S	S	S	S	S	S	T	T	S	T
Foramsulfuron	S	S	S	S	S	S	T	S	S	T
Glyphosate	S	S	S	S	S	D	D	S	S	S
Imazapic	S	S	S-I	S-I	S	S-I	I	T	S	I
Imazaquin	S	S	S	S	S	S	T	T	T	T
Mesotrione	S	T	T	T	T	S	S	T	T	S
Metsulfuron	S-I	I	S-I	I	S	S	T	I	T	T
Metribuzin	S	S	S	S	S	S	T	S	S	S
Pronamide	S	S	S	S	S	S	T	S	S	S
Quinclorac	I	T	T	I	T	S	T**	S	S	T
Rimsulfuron	S	S	S	S	S	S	T	S	S	T
Sethoxydim	S	S	S	T	S	S	S	T	S	S
Sulfosulfuron	S	S	S	S	S	S	T	T	T	T
Topramezone	S	T	T	T	T	S	S	T	S	S
Trifloxysulfuron	S	S	S	S	S	S	T	S	S	T

KEY: I = intermediate tolerance, use with caution, use at reduced label rates or minimum label rates; S = sensitive, DO NOT USE THIS HERBICIDE; T = tolerant. Since tolerance frequently depends upon rate used and formulation selected, CAREFULLY CHECK THE LABEL.

* Use only on Tifway 419 bermudagrass.

**Hybrid bermudagrass is intermediately tolerant to quinclorac.

Susceptibility of Broadleaf Weeds to Postemergence Turf Herbicides

Weed	Classification of Weed	2,4-D	Mecoprop	Dicamba	Triclopyr + Clopyralid	2,4-D + Triclopyr	2,4-D + Mecoprop	2,4-D + Dichlorprop	2,4-D + Mecoprop + Dicamba
					Response of Weeds to Herbicides				
Bittercress, hairy	WA	S	I	S	Unknown	Unknown	S	S	S
Buttercups	WA, B, P	S-I	I	I-R	S	Unknown	S	S	S
Buttonweed, Virginia	P	I	I-R	I-R	I	I	I	I	I
Carpetweed	SA	S	I	S	Unknown	S	S	S	S
Chickweed, common	WA	R	S-I	Unknown	S	S	S	S	S
Chickweed, mousear	WA, P	I-R	S-I	S	S	S	S	S	S
Chickory	P	S	S	S	Unknown	Unknown	S	S	S
Clover, hop	WA	I	S	S	S	S	S	S	S
Clover, white	P	I	S	S	S	S	S	S	S
Dandelion, catsear	P	S-I	I	S	S	S	S	S	S
Dandelion, common	P	S	S	S	Unknown	S	S	S	S
Dichondra	P	S	I	S-I	S	Unknown	S	S	S
Dock (broadleaf, curly)	P	I	I-R	S	S	S-I	I	I	S-I
Dogfennel	P	R	Unknown	S	Unknown	Unknown	Unknown	Unknown	Unknown
Garlic, wild	P	S-I	R	S-I	S	Unknown	S-I	S-I	S-I
Geranium, Carolina	WA	S	S-I	S	I	Unknown	S	S	S
Hawkweed	P	S-I	R	S-I	Unknown	Unknown	S-I	S-I	S-I
Healall	P	S	R	S-I	Unknown	Unknown	S	S	S
Henbit	WA	I-R	I	S	S	S	I	S-I	S
Horseweed	WA, SA	I	Unknown	S	S	Unknown	Unknown	Unknown	S-I
Ivy, ground	P	I-R	I	S-I	S-I	S-I	I	I	S-I
Knawel	WA	R	I	S	Unknown	S	S-I	S-I	S
Knotweed, prostrate	SA	R	I	S	Unknown	Unknown	S-I	S-I	S
Lespedeza	SA	I-R	S	S	S	W	S-I	I	S
Mallow	SA	I-R	I	S-I	Unknown	S	S-I	S-I	S-I
Medic, black	A	R	I	S	S	S	I	S	S
Mugwort	P	I	I-R	S-I	Unknown	Unknown	I	I	I
Parsley-piert	WA	R	S-I	S-I	Unknown	S	S-I	R	S-I
Pennywort, lawn	P	S-I	S-I	S-I	Unknown	Unknown	S-I	S-I	S-I
Plantains	P	S	I-R	R	S	S	S	S	S
Purslane, common	SA	I	R	S	Unknown	S	I	I	S-I
Sorrel, red	P	R	S	S	S	Unknown	S-I	I	S
Speedwell, corn	WA	I-R	I-R	I-R	Unknown	Unknown	I-R	I-R	I-R
Spurge, prostrate	SA	I	I	S	S-I	S-I	I	S-I	S-I
Spurge, spotted	SA	I-R	S-I	S-I	S-I	S-I	S-I	S-I	S
Spurweed (lawn burweed)	WA	I	S-I	S	S	S	S-I	I	S
Strawberry, India mock	P	R	I	S-I	Unknown	Unknown	I	R	S-I
Vetch, common	WA, SA	S	S	S	S	S	S	S	S
Violet, Johnny jump-up, wild pansy	WA	I-R	I-R	S-I	S-I	I	I-R	I	I-R
Violet, wild	P	I-R	I-R	S-I	S-I	S-I	I-R	I	I-R
Woodsorrel (common yellow)	P	R	R	I	S-I	S-I	I-R	I-R	I-R
Yarrow	P	I	I-R	S	Unknown	Unknown	I-R	I	S-I
Yellow rocket	WA	S-I	I	S-I	Unknown	Unknown	S-I	S-I	S

KEY: A = annual; B = biennial; P = perennial; SA = summer annual; WA = winter annual; S = susceptible; I = intermediately susceptible, good control can sometimes be achieved with high rates, but a repeat treatment 3 to 4 weeks later, each at the standard or reduced rate, is usually more effective; R = resistant in most cases.

Trade Names for Selected Postemergence Broadleaf Herbicides

2,4-D
Dymec
Lesco A-4D
Weedar 64
Weedestroy AM-40 Amine Salt

2,4-D + Dicamba
Lesco Eight-One
81 Selective Weedkiller

2,4-D + Dichlorprop (2,4-DP)
Turf D + DP amine
Turf 2D + 2DP ester
Weedone DPC

2,4-D + Mecoprop (or MCPP)
Ortho Weed-B-Gon Weed Killer*
Phenomec
2 MCPP + 2D Amine Turf Herbicide
2 Plus 2
2,4-D + Triclopyr
Chaser

2,4-D + Mecoprop + Dichlorprop
Weedestroy Triamine
Weedestroy Triester
Spoiler

2,4-D + Mecoprop + Dicamba
Lesco Three-Way
MEC Amine-D
MEC Amine-BG
33 Plus*
Trimec Classic, Trimec Southern
Trimec Bentgrass, Trimec 1000
Triplet SF
Weed-B-Gone for Southern Lawns
Formula II*

Atrazine
AAtrex 4L, AAtrex Nine-O
Atrazine 4L, Atrazine DF
Bonus S*

Bentazon + Arazine
Prompt

Chlorsulfuron
Corsair

Dicamba
Banvel
Vanquish

Diquat
Reward

Glufosinate
Finale

Glyphosate
Roundup Pro
Roundup ProDry
Roundup*
Ortho Kleenup*
Lesco Avail
Weed Wrangler

Imazapic
Plateau DG

MCPA + Mecoprop + Dicamba
Lesco Eliminate
Trimec Encore
Tri-Power

MCPA + Mecoprop + Dichlorprop
Weedestroy Triamine II
Weedestroy Triester II

MCPA + Triclopyr + Dicamba
Horsepower (amine)
Cool Power (ester)

MCPA + Fluroxypyr + Dicamba
Change Up

MCPA + Fluroxypyr + Triclopyr
Battleship III

Mecoprop (MCPP)
MCPP-p 4 Amine
Mecomec 4, Mecomec 2.5
Ortho Chickweed & Clover Control*
Weedestroy MCPP-4 Amine

Metsulfuron
Blade
Manor

MSMA
MSMA Turf
MSMA 6.6

MSMA + 2,4-D + Mecoprop + Dicamba
Trimec Plus (Quadmec)

Simazine
Princep Liquid, Princep DF
Regal Wynstar
Sim-trol

Triclopyr + Clopyralid
Confront

* Products available for homeowner use.

Annual Grassy Weed Control Ratings for Turf Herbicides

Herbicide	Crabgrass	Goosegrass	Annual Bluegrass
Benefin	Good	Fair	Good
Bensulide	Good	Poor	Good
Clethodim	Good to Excellent	Good to Excellent	Good to Excellent
DCPA	Good	Poor	Good
Dimethenamid	Poor	Good to Excellent	Poor
Dithiopyr	Good to Excellent	Good	Good
Indaziflam	Good to Excellent	Good to Excellent	Good to Excellent
Metolachlor	Good	Fair	Good
Napropamide	Good	Good	Good
Oryzalin	Good to Excellent	Good	Good
Oxadiazon	Good	Good	Good
Pendimethalin	Good to Excellent	Fair to Good	Good
Prodiamine	Good to Excellent	Good	Good
Quinclorac	Excellent	Not Registered	Not Registered
Siduron	Good	Fair	Not Registered

Excellent = 90 to 100% effective control; Good = 80 to 90% effective control; Fair = 70 to 80% effective control; Poor = less than 70% effective control.

Turfgrass Disease Control

J. P. Kerns and E. L. Butler, Entomology and Plant Pathology Extension

When more than one brand name exists for an agricultural chemical, the brand name that first came onto the market is listed first. Otherwise, brand names are listed in alphabetical order. The order in which brand names are given is not an indication of a recommendation or criticism. Products marked with an asterisk (*) are not labeled for home lawn use.

Turfgrass Disease Control

Disease	Fungicide and Formulation[1]	Amount of Formulation (oz/1,000 sq ft)[2]	Application Interval (days)[3]
Algae	benzovindiflupyr + difenoconazole (Ascernity) 0.86 SL*	1	14
(*Cyanobacteria*)	boscalid + chlorothalonil (Encartis) 6.25 SC*	3 to 4	14
	chlorothalonil*		
	(Daconil) 82.5 WDG	1.8 to 3.25	7 to 14
	(Daconil Weather Stik, Legend) 6 F	2 to 3.6	7 to 14
		4 to 5.5	14
	(Daconil Zn) 4.16 F	3 to 5	7 to 14
		6 to 8	14
	chlorothalonil + acibenzolar-S-methyl (Daconil Action) 6.1 F*	2 to 3.6	7 to 14
		4 to 5.4	14
	chlorothalonil + azoxystrobin (Renown) 5.16 SC*	2.5 to 4.5	10 to 14
	chlorothalonil + fluoxastrobin (Fame C) 4.25 SC*	3 to 5.4	7 to 14
	chlorothalonil + thiophanate-methyl (Spectro) 90 WDG*	2 to 5.76	7 to 14
	chlorothalonil + triticonazole (Reserve) 4.79 SC*	3.2 to 5.4	14 to 28
	fluazinam (Secure) 4.17 SC*	0.5	14
	fluazinam + acibenzolar-S-methyl (Secure Action) 4.18 SC*	0.5	14
	fluazinam + tebuconazole (Traction) 3.24 SC*	1.3	14
	fluxapyroxad (Xzemplar) 2.47 SC	0.21 to 0.26	14 to 28
	mancozeb*		
	(Fore) 80 WP	6	7 to 14
	(Dithane, Pentathlon) 75 DF	6	refer to label
	(Pentathlon) 4 LF	10	refer to label
	(Protect) 75 WP	6	7 to 14
	(Wingman) 75 WP	6	refer to label
	mancozeb + copper hydroxide (Junction) 60 DF*	2 to 4	7 to 14
Anthracnose	azoxystrobin		
(*Colletotrichum cereale*)	(Heritage, Strobe) 50 WG	0.2 to 0.4	14 to 28
	(Heritage) 0.8 TL	1 to 2	14 to 28
	(Heritage) 0.31 G	2 to 4 lbs	14 to 28
	(Strobe) 2 L	0.38 to 0.77	14 to 28
	azoxystrobin + acibenzolar-S-methyl (Heritage Action) 51 WG*	0.2 to 0.4	14 to 28
	azoxystrobin + chlorothalonil (Renown) 5.16 SC*	2.5	7 to 10
		4.5	14 to 21
	azoxystrobin + difenoconazole (Briskway) 2.7 SC*	0.5 to 1.2	14
	azoxystrobin + propiconazole (Headway)		
	1.4 ME	1.5 to 3	14 to 28
	1.06 G	2 to 4 lbs	14 to 28
	azoxystrobin + tebuconazole (Strobe T) 2.67 SC*	0.75 to 1.5	14 to 21
	benzovindiflupyr + difenoconazole (Ascernity) 0.86 SL*	1	14
	boscalid + chlorothalonil (Encartis) 6.25 SC*	3 to 4	14

Turfgrass Disease Control (continued)

Disease	Fungicide and Formulation[1]	Amount of Formulation (oz/1,000 sq ft)[2]	Application Interval (days)[3]
Anthracnose (*Colletotrichum cereale*) (continued)	chlorothalonil*		
	(Daconil Ultrex) 82.5 WDG	2.75 to 5	7 to 14
	(Daconil Weather Stik, Legend) 6 F	3 to 3.6	7 to 14
		3.6 to 5.5	14
	(Daconil Zn) 4.16 F	4.4 to 5	7 to 14
		5.3 to 8	14
	(Chlorothalonil 500ZN) 4.17 F	3 to 5	7 to 14
		7.9	14
	(Chlorothalonil 720SFT) 6 F	2.12 to 3.5	7 to 14
		5.5	14
	(Chlorothalonil, Chlorostar) 82.5 DF	2.8 to 3.2	7 to 14
	(Pegasus) 6 L	3.6 to 5.5	7 to 14
	(Pegasus) 82.5 DF	3.25 to 5	7 to 14
	(Pegasus HPX) 6 F	3.6 to 5.5	7 to 14
	chlorothalonil + acibenzolar-S-methyl (Daconil Action) 6.1 F*	3 to 3.6	7 to 14
		3.6 to 5.4	14
	chlorothalonil + fluoxastrobin (Fame C) 4.25 SC*	3 to 5.9	14 to 28
	chlorothalonil + iprodione + thiophanate-methyl + tebuconazole (Enclave) 5.3 F*	3 to 4	14 to 21
		7 to 8	28
	chlorothalonil + propiconazole (Concert) 4.3 SC*	4.5 to 8.5	7 to 28
	chlorothalonil + propiconazole + fludioxonil (Instrata) 3.59 SC*	2.75 to 6	14 to 28
	chlorothalonil + thiophanate-methyl*		
	(Consyst) 67 WDG	2 to 8	7 to 14
	(Peregrine) 67 WDG	2 to 8	14
	(Spectro) 90 WDG	3.72 to 5.76	7 to 14
	(TM/C) 67 WDG	2 to 8	14 to 21
	cyazofamid + azoxystrobin (Union) 0.79 SC	2.9 to 5.75	14 to 28
	fenarimol (Rubigan) 1 AS*	1.75 to 3.5	30
	fluazinam (Secure) 4.17 SC*	0.5	14
	fluazinam + acibenzolar-S-methyl (Secure Action) 4.18 SC*	0.5	14
	fluazinam + tebuconazole (Traction) 3.24 SC*	1.3	14
	fludioxonil (Medallion) 50 WP	0.25 to 0.5	14
	fluopyram + trifloxystrobin (Exteris Stressgard) 0.27 SC	2.135 to 6	14 to 28
	fluoxastrobin (Fame)		
	4 SC	0.18 to 0.36	14 to 28
	0.25 G	2.3 to 4.6 lbs	14 to 28
	fluoxastrobin + myclobutanil (Fame M) 3.9 SC	0.25 to 1	14 to 28
	fluoxastrobin + tebuconazole (Fame T) 4 SC*	0.45 to 0.9	21 to 28
	flutolanil + thiophanate-methyl (SysStar) 80 WDG	2 to 3	14 to 30
	flutriafol (Rayora) 1.04 L*	0.7 to 1.4	14 to 21
	iprodione + thiophanate-methyl*		
	(26/36) 3.8 F	2 to 4	14 to 21
	(Dovetail) 3.8 F	1 to 4	14 to 21
	iprodione + trifloxystrobin (Interface) 2.27 SC*	4 to 7	refer to label
	isofetamid + tebuconazole (Tekken) 1.8 SC*	3	14 to 28
	mefentrifluconazole (Maxtima) 3.34 SC*	0.4 to 0.6	14
	mefentrifluconazole + pyraclostrobin (Navicon) 3.34 SC*	0.7 to 0.85	14 to 28
	metconazole (Tourney) 50 WDG	0.28 to 0.37	14 to 21
	mineral oil (Civitas) + proprietary pigment (Civitas Harmonizer)*	(8 to 32) + (1 to 4)	7 to 21
	myclobutanil (Eagle, Myclobutanil, Siskin) 20 EW	1.2	14 to 21
	penthiopyrad (Velista) 50 WG	0.3 to 0.5	14
	phosphorous acid (Jetphiter) 5.41 F	5	7
	polyoxin D		
	(Affirm) 11.3 WDG	0.88	7 to 14
	(Endorse) 2.5 WP	4	7 to 14

Turfgrass Disease Control (continued)

Disease	Fungicide and Formulation[1]	Amount of Formulation (oz/1,000 sq ft)[2]	Application Interval (days)[3]
Anthracnose (*Colletotrichum cereale*) (continued)	propiconazole (Banner MAXX, Kestrel, Propiconazole, Savvi, Strider) 1 ME	1 to 2	14 to 28
	prothioconazole (Densicor) 4 SC*	0.196	14 to 21
	Pseudomonas chlororaphis strain AFS009 (Zio) SC	1.8 to 6.0	7 to 21
	pydiflumetofen + azoxystrobin + propiconazole (Posterity XT) 1.48 SE*	1.5 to 3	14 to 28
	pyraclostrobin (Insignia)		
	20 WG	0.5 to 0.9	14 to 28
	2 SC	0.4 to 0.7	14 to 28
	pyraclostrobin + boscalid (Honor) 28 WG*	0.55 to 1.1	14 to 28
	pyraclostrobin + fluxapyroxad (Lexicon Intrinsic) 4.17 SC	0.34 to 0.47	14 to 28
	pyraclostrobin + triticonazole (Pillar) 0.81 G	3 lbs	14 to 28
	tebuconazole*		
	(Torque) 3.6 F	0.6 to 1.1	21
	(Mirage, Stressgard) 2 SC	1 to 2	14 to 28
	(Skylark, Tebuconazole) 3.6 F	0.6	28
	thiophanate-methyl		
	(3336) 50 WP or 4 F	2 to 6	14
	(3336 Plus) 2 F	2 to 8	14 to 28
	(SysTec 1998, T-Bird, TM) 85 WDG	0.67 to 1.3	14
	(3336) 2 G	3 to 9 lbs	14
	(SysTec 1998, T-Bird, TM) 4.5 L	1 to 2	14
	triadimefon (Bayleton) 50 WSP	1	30 to 45
	trifloxystrobin (Compass) 50 WDG	0.15 to 0.25	14 to 21
	trifloxystrobin + triadimefon		
	(Armada) 50 WP	0.6 to 1.2	14 to 28
	(Tartan) 2 SC*	1 to 2	14 to 28
	triticonazole		
	(Trinity) 1.7 SC	0.5 to 1	14 to 28
	(Triton) 70 WDG	0.15 to 0.225	14 to 28
	(Triton Flo) 3 F	0.41 to 1.1	14 to 28
	triticonazole + chlorothalonil (Reserve) 4.79 SC*	3.2 to 5.4	14 to 28
Brown Patch (*Rhizoctonia solani*)	azoxystrobin		
	(Heritage, Strobe) 50 WG	0.2 to 0.4	14 to 28
	(Heritage) 0.8 TL	1 to 2	14 to 28
	(Heritage) 0.31 G	2 to 4 lbs	14 to 28
	(Strobe) 2 L	0.38 to 0.77	14 to 28
	azoxystrobin + acibenzolar-S-methyl (Heritage Action) 51 WG*	0.2 to 0.4	14 to 28
	azoxystrobin + chlorothalonil (Renown) 5.16 SC*	2.5	14
		4.5	14 to 21
	azoxystrobin + difenoconazole (Briskway) 2.7 SC*	0.5 to 1.2	14 to 28
	azoxystrobin + propiconazole (Headway)		
	1.4 ME	0.75 to 3	14 to 28
	1.06 G	2 to 4 lbs	14 to 28
	azoxystrobin + tebuconazole (Strobe T) 2.67 SC*	0.75 to 1.5	14 to 21
	benzovindiflupyr + difenoconazole (Ascernity) 0.86 SL*	1	14 to 21
	boscalid + chlorothalonil (Encartis) 6.25 SC*	3 to 4	14
	chloroneb (Teremec)*		
	65 SP	3 to 4	7 to 10
	2.9 F	5 to 7	7 to 10

Turfgrass Disease Control (continued)

Disease	Fungicide and Formulation[1]	Amount of Formulation (oz/1,000 sq ft)[2]	Application Interval (days)[3]
Brown Patch	chlorothalonil*		
(*Rhizoctonia solani*)	(Daconil Ultrex) 82.5 WDG	1.8 to 3.23	7 to 14
(continued)		3.7 to 5	14
	(Daconil Weather Stik, Legend) 6 F	2 to 3.6	7 to 14
		4 to 5.5	14
	(Daconil Zn) 4.16 F	3 to 5	7 to 14
		6 to 8	14
	(Chlorothalonil 500ZN) 4.17 F	3 to 5	7 to 14
		7.9	14
	(Chlorothalonil 720SFT) 6 F	2.12 to 3.5	7 to 14
		5.5	14
	(Chlorothalonil, Chlorostar) 82.5 DF	1.8 to 3.2	7 to 14
	(Pegasus) 6 L	2 to 3.6	7 to 14
	(Pegasus) 82.5 DF	1.82 to 3.25	7 to 14
	(Pegasus HPX) 6 F	2 to 3.6	7 to 14
	chlorothalonil + acibenzolar-S-methyl (Daconil Action) 6.1 F*	2 to 3.5	7 to 14
		4 to 5.4	14
	chlorothalonil + fluoxastrobin (Fame C) 4.25 SC*	1.5 to 5.9	14 to 28
	chlorothalonil + iprodione + thiophanate-methyl + tebuconazole (Enclave) 5.3 F*	3 to 4	14 to 21
		7 to 8	28
	chlorothalonil + propiconazole (Concert) 4.3 SC*	3 to 5.5	7 to 14
		5.5 to 8.5	14 to 28
	chlorothalonil + propiconazole + fludioxonil (Instrata) 3.6 SC*	2.75 to 6	14 to 21
	chlorothalonil + thiophanate-methyl*		
	(Spectro) 90 WDG	3 to 5.76	14 to 21
	(TM/C) 67 WDG	2 to 8	14 to 21
	cyazofamid + azoxystrobin (Union) 0.79 SC	2.9 to 5.75	14 to 28
	fenarimol (Rubigan) 1 AS*	1.5	7 to 14
	fluazinam (Secure) 4.17 SC*	0.5	14
	fluazinam + acibenzolar-S-methyl (Secure Action) 4.18 SC*	0.5	14
	fluazinam + tebuconazole (Traction) 3.24 SC*	1.3	14
	fludioxonil (Medallion) 50 WP	0.2 to 0.25	7
		0.5	14
	fluopyram + trifloxystrobin (Exteris Stressgard) 0.27 SC	2.135 to 6	14 to 28
	fluoxastrobin (Fame)		
	4 SC	0.09 to 0.36	14 to 28
	0.25 G	1.2 to 4.6 lbs	14 to 28
	fluoxastrobin + myclobutanil (Fame M) 3.9 SC	0.25 to 1	14 to 28
	fluoxastrobin + tebuconazole (Fame T) 4 SC*	0.45 to 0.9	21 to 28
	flutriafol (Rayora) 1.04 L*	0.7 to 1.4	14 to 21
	fluxapyroxad (Xzemplar) 2.47 SC	0.21 to 0.26	14 to 21
	flutolanil		
	(Prostar) 70 WP, 70 DG	1.5 to 3	14 to 21
	(Pedigree) 3.8 SC	2.2 to 4.4	14 to 21
	flutolanil + thiophanate-methyl (SysStar) 80 WDG	2 to 3	14 to 21
	iprodione 26GT, Iprodione Pro, IPro, Raven* 2 F, 2 SC, 2 SE	3 to 4	14 to 28
	iprodione + thiophanate-methyl*		
	(26/36) 3.8 F	2 to 4	14 to 21
	(Dovetail) 3.8 F	1 to 4	14 to 21
	iprodione + trifloxystrobin (Interface) 2.27 SC*	3 to 5	refer to label
	isofetamid + tebuconazole (Tekken) 1.8 SC*	3	14 to 28
	mancozeb*		
	(Fore) 80 WP	4	7
	(Dithane) 75 DF	4	10
	(Protect) 75 WP	4	7 to 14

Turfgrass Disease Control (continued)

Disease	Fungicide and Formulation[1]	Amount of Formulation (oz/1,000 sq ft)[2]	Application Interval (days)[3]
Brown Patch	mancozeb + copper hydroxide (Junction) 60 DF*	2 to 4	7
(*Rhizoctonia solani*)	mandestrobin (Pinpoint) 4SC	0.31	14
(continued)	mefentrifluconazole + pyraclostrobin (Navicon) 3.34 SC*	0.7 to 0.85	14 to 28
	metconazole (Tourney) 50 WDG	0.28 to 0.37	14 to 21
	mineral oil (Civitas) + proprietary pigment (Civitas Harmonizer)*	(8 to 32) + (1 to 4)	7 to 21
	myclobutanil (Eagle, Myclobutanil, Siskin) 20 EW	1.2	14
	penthiopyrad (Velista) 50 WG	0.3 to 0.5	14 to 21
	phosphorous acid (Jetphiter) 5.41 F	5	7
	polyoxin D		
	(Affirm) 11.3 WDG	0.88	7 to 14
	(Endorse) 2.5 WP	4	7 to 14
	propiconazole (Banner MAXX, Kestrel, Propiconazole, Savvi, Strider) 1 ME	1 to 2	14 to 21
	prothioconazole (Densicor) 4 SC*	0.196	14 to 21
	Pseudomonas chlororaphis strain AFS009 (Zio) SC	1.8 to 6.0	7 to 21
	pydiflumetofen + azoxystrobin + propiconazole (Posterity Forte) 2.5 SE*	0.63 to 0.84	14 to 21
	pydiflumetofen + azoxystrobin + propiconazole (Posterity XT) 1.48 SE*	0.75 to 1.5	14
	pyraclostrobin (Insignia)		
	20 WG	0.5 to 0.9	14 to 28
	2 SC	0.4 to 0.7	14 to 28
	pyraclostrobin + boscalid (Honor) 28 WG*	0.55 to 1.1	14 to 28
	pyraclostrobin + fluxapyroxad (Lexicon Intrinsic) 4.17 SC	0.34 to 0.47	14 to 28
	pyraclostrobin + triticonazole (Pillar) 0.81 G	3 lbs	14 to 28
	tebuconazole*		
	(Torque) 3.6 F	0.6 to 1.1	21
	(Mirage, Stressgard) 2 SC	1 to 2	14 to 28
	(Skylark, Tebuconazole) 3.6 F	0.6	28
	thiram (Spotrete) 4 F*	3.75 to 7.5	3 to 10
	triadimefon (Bayleton) 50 WSP, 4.15 F	0.5 to 1	15 to 30
	trifloxystrobin (Compass) 50 WDG	0.1 to 0.2	14
		0.15 to 0.25	21
	trifloxystrobin + triadimefon		
	(Armada) 50 WP	0.6 to 1.2	14 to 28
	(Tartan) 2 SC*	1 to 2	14 to 28
	triticonazole		
	(Trinity) 1.7 SC	0.75 to 2	14 to 28
	(Triton) 70 WDG	0.15 to 0.3	14 to 28
	(Triton Flo) 3 F	0.41 to 1.1	14 to 28
	triticonazole + chlorothalonil (Reserve) 4.79 SC*	3.2 to 5.4	14 to 28
	vinclozolin (Curalan, Touche) 50 EG*	1	14 to 28
Brown Ring Patch	azoxystrobin (Heritage)		
(*Rhizoctonia circinata* var.	50 WG	0.2 to 0.4	14 to 28
circinata)	0.8 TL	1 to 2	14 to 28
	0.31 G	2 to 4 lbs	14 to 28
	azoxystrobin + acibenzolar-S-methyl (Heritage Action) 51 WG*	0.2 to 0.4	14 to 28
	azoxystrobin + difenoconazole (Briskway) 2.7 SC*	0.5 to 1.2	14 to 28
	azoxystrobin + propiconazole (Headway) 1.4 ME	1.5 to 3	14 to 28
	benzovindiflupyr + difenoconazole (Ascernity) 0.86 SL*	1	14 to 21
	fluazinam + tebuconazole (Traction) 3.24 SC*	1.3	21
	fluoxastrobin + myclobutanil (Fame M) 3.9 SC	0.25 to 1	14 to 28
	fluoxastrobin + tebuconazole (Fame T) 4 SC*	0.45 to 0.9	21 to 28
	isofetamid + tebuconazole (Tekken) 1.8 SC*	3	14 to 28
	penthiopyrad (Velista) 50 WG	0.5	14
	polyoxin D		
	(Affirm) 11.3 WDG	0.88	7 to 14
	(Endorse) 2.5 WP	4	7 to 14

Turfgrass Disease Control (continued)

Disease	Fungicide and Formulation[1]	Amount of Formulation (oz/1,000 sq ft)[2]	Application Interval (days)[3]
Brown Ring Patch (*Rhizoctonia circinata var. circinata*) (continued)	prothioconazole (Densicor) 4 SC*	0.196	14 to 21
	pydiflumetofen + azoxystrobin + propiconazole (Posterity XT) 1.48 SE*	1.5 to 3	14 to 28
	pyraclostrobin (Insignia) 2 SC	0.7	14 to 28
	pyraclostrobin + fluxapyroxad (Lexicon Intrinsic) 4.17 SC	0.34 to 0.47	14 to 28
	pyraclostrobin + triticonazole (Pillar) 0.81 G	3 lbs	14 to 28
	tebuconazole*		
	(Torque) 3.6 F	0.6 to 1.1	21
	(Mirage, Stressgard) 2 SC	1 to 2	14 to 28
	(Skylark, Tebuconazole) 3.6 F	0.6	28
	triticonazole		
	(Trinity) 1.7 SC	1 to 2	14 to 28
	(Triton FLO) 3 F	0.5 to 1.1	14 to 28
	triticonazole + chlorothalonil (Reserve) 4.79 SC*	3.2 to 5.4	14 to 28
Copper Spot (*Gloeocercospora sorghi*)	benzovindiflupyr + difenoconazole (Ascernity) 0.86 SL*	1	14
	boscalid + chlorothalonil (Encartis) 6.25 SC*	4	14
	chlorothalonil*		
	(Daconil Ultrex) 82.5 WDG	3.7 to 5	14
	(Daconil Weather Stik, Legend) 6 F	4 to 5.5	14
	(Daconil Zn) 4.16 F	6 to 8	14
	(Chlorothalonil 500ZN) 6 F	3 to 5	7 to 10
		7.9	14
	(Chlorothalonil 720SFT) 6 F	2.12 to 3.5	7 to 10
		5.5	14
	(Chlorothalonil, Chlorostar) 82.5 DF	3.2	7 to 10
	(Pegasus) 6 L	3.6 to 5.5	7 to 14
	(Pegasus) 82.5 DF	3.25 to 5	7 to 14
	(Pegasus HPX) 6 F	3.6 to 5.5	7 to 14
	chlorothalonil + acibenzolar-S-methyl (Daconil Action) 6.1 F*	4 to 5.4	14
	chlorothalonil + azoxystrobin (Renown) 5.16 SC*	2.5	14
	chlorothalonil + fluoxastrobin (Fame C) 4.25 SC*	5.9	14
	chlorothalonil + iprodione + thiophanate-methyl + tebuconazole	3 to 4	14 to 21
	(Enclave) 5.3 F*	7 to 8	28
	chlorothalonil + propiconazole (Concert) 4.3 SC*	5.5 to 8.5	14
	chlorothalonil + thiophanate-methyl*		
	(Consyst) 67 WDG	3 to 8	7 to 10
	(Peregrine) 67 WDG	3 to 8	14
	(Spectro) 90 WDG	3 to 5.76	14
	(TM/C) 67 WDG	3 to 8	14 to 21
	fenarimol (Rubigan) 1 AS*	0.75 to 1.5	10 to 28
	fluazinam + tebuconazole (Traction) 3.24 SC*	1.3	14
	fluoxastrobin + myclobutanil (Fame M) 3.9 SC	0.25 to 1	14 to 21
	flutolanil + thiophanate-methyl (SysStar) 80 WDG	2 to 3	14 to 21
	flutriafol (Rayora) 1.04 L*	0.7 to 1.4	14 to 21
	iprodione + thiophanate-methyl (26/36) 3.8 F*	2 to 4	14 to 21
	isofetamid + tebuconazole (Tekken) 1.8 SC*	3	14 to 28
	mancozeb*		
	(Fore) 80 WP	4 to 8	7 to 14
	(Dithane) 75 DF	4 to 8	10
	(Pentathlon) 4 LF	7 to 14	7 to 14
	(Pentathlon) 75 DF	4 to 8	7
	(Protect, Wingman) 75 WP	4 to 8	7 to 14
	mancozeb + copper hydroxide (Junction) 60 DF*	2 to 4	7 to 14
	myclobutanil (Eagle, Myclobutanil, Siskin) 20 EW	1.2	14
	tebuconazole*		
	(Torque) 3.6 F	0.6 to 1.1	refer to label
	(Skylark, Tebuconazole) 3.6 F	0.6	28

Turfgrass Disease Control (continued)

Disease	Fungicide and Formulation[1]	Amount of Formulation (oz/1,000 sq ft)[2]	Application Interval (days)[3]
Copper Spot (*Gloeocercospora sorghi*) (continued)	thiophanate-methyl		
	(3336) 50 WP or 4 F	2 to 4	14
	(3336 Plus) 2 F	2 to 4	14 to 28
	(SysTec 1998, T-Bird, TM) 85 WDG	0.67 to 1.3	14
	(3336) 2 G	1.5 to 6 lbs	14
	(SysTec 1998, T-Bird, TM) 4.5 L	1 to 2	14
	thiram (Spotrete) 4F*	3.75 to 7.5	3 to 10
	triadimefon (Bayleton) 50 WSP, 4.15 F	0.5 to 1	15 to 30
Dead Spot (*Ophiosphaerella agrostis*)	azoxystrobin + propiconazole (Headway)		
	1.4 ME	1.5 to 3	14
	1.06 G	2 to 4 lbs	14 to 28
	azoxystrobin + tebuconazole (Strobe T) 2.67 SC*	0.75 to 1.5	14
	boscalid* (Emerald) 70 WG	0.18	14
	boscalid + chlorothalonil (Encartis) 6.25 SC*	4	14
	chlorothalonil + thiophanate-methyl (Spectro) 90 WDG*	3.72 to 5.76	14
	fludioxonil (Medallion) 50 WP	0.3 to 0.5	14
	pydiflumetofen + azoxystrobin + propiconazole (Posterity XT) 1.48 SE*	1.5 to 3	14
	pyraclostrobin (Insignia)		
	20 WG	0.5 to 0.9	14 to 28
	2 SC	0.4 to 0.7	14 to 28
	pyraclostrobin + boscalid (Honor) 28 WG*	0.55 to 1.1	14 to 28
	pyraclostrobin + fluxapyroxad (Lexicon Intrinsic) 4.17 SC	0.34 to 0.47	14 to 28
	pyraclostrobin + triticonazole (Pillar) 0.81 G	3 lbs	14 to 28
	thiophanate-methyl		
	(3336) 50WP or 4 F	4 to 6	14
	(3336 Plus) 2 F	4 to 6	14
	(3336) 2 G	6 to 9 lbs	14
Dollar Spot (*Clarireedia spp*)	azoxystrobin + difenoconazole (Briskway) 2.7 SC*	0.5 to 1.2	14 to 21
	azoxystrobin + propiconazole (Headway)		
	1.4 ME	0.75 to 3	7 to 28
	1.06 G	2 to 4 lbs	14 to 28
	azoxystrobin + tebuconazole (Strobe T) 2.67 SC*	0.75 to 1.5	14 to 21
	benzovindiflupyr + difenoconazole (Ascernity) 0.86 SL*	1	14 to 21
	boscalid* (Emerald) 70 WG	0.13 to 0.18	14 to 28
	boscalid + chlorothalonil (Encartis) 6.25 SC*	3 to 4	14 to 28
	chlorothalonil*		
	(Daconil Ultrex) 82.5W DG	1 to 3.25	7 to 21
		3.7 to 5	14 to 21
	(Daconil Weather Stik, Legend) 6 F	1 to 3.6	7 to 21
		4 to 5.5	14 to 21
	(Daconil Zn) 4.16 F	1.5 to 5	7 to 21
		6 to 8	14
	(Chlorothalonil 500ZN) 4.17 F	3 to 5	7 to 14
		7.9	14
	(Chlorothalonil 720SFT) 6 F	2.12 to 3.5	7 to 14
		5.5	14
	(Chlorothalonil, Chlorostar) 82.5 DF	1.8 to 3.2	7 to 10
	(Pegasus) 6 L	2 to 3.6	7 to 14
	(Pegasus) 82.5 DF	1.82 to 3.25	7 to 14
	(Pegasus HPX) 6 F	2 to 3.6	7 to 14
	chlorothalonil + acibenzolar-S-methyl (Daconil Action) 6.1 F*	1 to 3.5	7 to 21
		4 to 5.4	14
	chlorothalonil + azoxystrobin (Renown) 5.16 SC*	2.5 to 4.5	7 to 14
	chlorothalonil + fluoxastrobin (Fame C) 4.25 SC*	3 to 5.9	14 to 21

Turfgrass Disease Control (continued)

Disease	Fungicide and Formulation[1]	Amount of Formulation (oz/1,000 sq ft)[2]	Application Interval (days)[3]
Dollar Spot	chlorothalonil + iprodione + thiophanate-methyl + tebuconazole	3 to 4	14 to 21
(*Clarireedia spp*)	(Enclave) 5.3 F*	7 to 8	28
(continued)	chlorothalonil + propiconazole (Concert) 4.3 SC*	1.5 to 3	7 to 10
		3 to 5.5	14 to 21
		5.5 to 8.5	14 to 28
	chlorothalonil + propiconazole + fludioxonil (Instrata) 3.6 SC*	2.75 to 6	21 to 28
	chlorothalonil + thiophanate-methyl*		
	(Consyst) 67 WDG	2 to 8	7 to 21
	(Peregrine) 67 WDG	2 to 8	14
	(Spectro) 90 WDG	3.72 to 5.76	14 to 21
	(TM/C) 67 WDG	2 to 8	7 to 14
	fenarimol (Rubigan) 1 AS*	0.75 to 1.5	10 to 28
	fluazinam (Secure) 4.17 SC*	0.5	14
	fluazinam + acibenzolar-S-methyl (Secure Action) 4.18 SC*	0.5	14 to 21
	fluazinam + tebuconazole (Traction) 3.24 SC*	1.3	14
	fluopyram + trifloxystrobin (Exteris Stressgard) 0.27 SC	1.5 to 4.135	7 to 28
	fluoxastrobin (Fame)		
	4 SC	0.18 to 0.36	14 to 21
	0.25 G	2.3 to 4.6 lbs	14 to 21
	fluoxastrobin + myclobutanil (Fame M) 3.9 SC	0.25 to 1	14 to 21
	fluoxastrobin + tebuconazole (Fame T) 4 SC*	0.45 to 0.9	21 to 28
	flutriafol (Rayora) 1.04 L*	0.7 to 1.4	14 to 21
	fluxapyroxad (Xzemplar) 2.47 SC	0.16 to 0.26	14 to 28
	flutolanil + thiophanate-methyl (SysStar) 80 WDG	2 to 3	14 to 30
	iprodione (26GT, Iprodione Pro, IPro, Raven) 2 F, 2 SC, 2 SE*	2 to 4	14 to 28
	iprodione + thiophanate-methyl*		
	(26/36) 3.8 F	2 to 4	14 to 21
	(Dovetail) 3.8 F	1 to 4	14 to 21
	iprodione + trifloxystrobin (Interface) 2.27 SC*	2 to 5	refer to label
	isofetamid (Kabuto 3.33 SC)	0.4 to 0.5	14
	isofetamid + tebuconazole (Tekken) 1.8 SC*	3	14 to 28
	mancozeb*		
	(Fore) 80 WP	6 to 8	7 to 14
	(Dithane) 75 DF	6 to 8	10
	(Pentathlon) 4 LF	10 to 14	7 to 14
	(Pentathlon) 75 DF	6 to 8	7
	(Protect, Wingman) 75 WP	6 to 8	7 to 14
	mancozeb + copper hydroxide (Junction) 60 DF*	2 to 4	7 to 14
	mandestrobin (Pinpoint) 4SC	0.17 to 0.31	14 to 21
	mefentrifluconazole (Maxtima) 3.34 SC*	0.2 to 0.4	14 to 28
	mefentrifluconazole + pyraclostrobin (Navicon) 3.34 SC*	0.7 to 0.85	14 to 28
	metconazole (Tourney) 50 WDG	0.18 to 0.37	14 to 21
	mineral oil (Civitas) + proprietary pigment (Civitas Harmonizer)*	(8 to 32) + (1 to 4)	7 to 21
	myclobutanil (Eagle, Myclobutanil, Siskin) 20 EW	0.5 to 2.4	7 to 28
	penthiopyrad (Velista) 50 WG	0.3 to 0.5	14 to 21
	propiconazole (Banner MAXX, Propiconazole, Savvi, Spectator) 1 ME	0.5 to 2	7 to 28
	prothioconazole (Densicor) 4 SC*	0.196	14 to 21
	Pseudomonas chlororaphis strain AFS009 (Zio) SC	1.8 to 6.0	7 to 21
	pydiflumetofen (Posterity) 1.67 SC*	0.08 to 0.32	14 to 28
	pydiflumetofen + azoxystrobin + propiconazole (Posterity Forte) 2.5 SE*	0.42 to 0.84	21 to 28
	pydiflumetofen + azoxystrobin + propiconazole (Posterity XT) 1.48 SE*	1.5 to 3	14 to 28
	pyraclostrobin (Insignia)		
	20 WG	0.9	14
	2 SC	0.7	14
	pyraclostrobin + boscalid (Honor) 28 WG*	0.83 to 1.1	14 to 21
	pyraclostrobin + fluxapyroxad (Lexicon Intrinsic) 4.17 SC	0.34 to 0.47	14 to 28

Turfgrass Disease Control (continued)

Disease	Fungicide and Formulation[1]	Amount of Formulation (oz/1,000 sq ft)[2]	Application Interval (days)[3]
Dollar Spot	pyraclostrobin + triticonazole (Pillar) 0.81 G	3 lbs	14 to 28
(*Clarireedia spp*)	tebuconazole*		
(continued)	(Torque) 3.6 F	0.6 to 1.1	refer to label
	(Mirage Stressgard) 2 SC	1 to 2	14 to 28
	(Skylark, Tebuconazole) 3.6 F	0.6	28
	thiophanate-methyl		
	(3336) 50WP or 4 F	2 to 4	14
	(3336 Plus) 2 F	2 to 4	14 to 28
	(SysTec 1998, T-Bird, TM) 85 WDG	0.67 to 1.3	14
	(3336) 2 G	1.5 to 6 lbs	14
	(SysTec 1998, T-Bird, TM) 4.5 L	1 to 2	14
	thiram (Spotrete) 4 F*	3.75 to 7.5	3 to 10
	triadimefon (Bayleton) 50 WSP, 4.15 F	0.25 to 1	14 to 30
	trifloxystrobin + triadimefon		
	(Armada) 50 WP	0.6 to 1.2	14 to 28
	(Tartan) 2 SC*	1 to 2	14 to 28
	triticonazole		
	(Trinity) 1.7 SC	1 to 2	14 to 28
	(Triton) 70 WDG	0.15 to 0.3	14 to 28
	(Triton FLO) 3 F	0.28 to 1.1	14 to 28
	triticonazole + chlorothalonil (Reserve) 4.79 SC*	3.2 to 4.5	14 to 28
	vinclozolin (Curalan, Touche) 50 EG*	1	21 to 28
Fairy Ring	azoxystrobin		
(*Basidiomycetes*)	(Heritage, Strobe) 50 WG	0.4	28
	(Heritage) 0.8 TL	2	28
	(Heritage) 0.31 G	2 to 4 lbs	14 to 28
	azoxystrobin + acibenzolar-S-methyl (Heritage Action) 51 WG*	0.2 to 0.4	14 to 28
	azoxystrobin + difenoconazole (Briskway) 2.7 SC*	0.5 to 1.2	14 to 28
	azoxystrobin + propiconazole (Headway)		
	1.4ME	1.5 to 3	14 to 28
	1.06 G	2 to 4 lbs	14 to 28
	azoxystrobin + tebuconazole (Strobe T) 2.67 SC*	0.75 to 1.5	28
	benzovindiflupyr + difenoconazole (Ascernity) 0.86 SL*	1	14 to 28
	chlorothalonil + fluoxastrobin (Fame C) 4.25 SC*	4.5 to 5.9	21 to 28
	cyazofamid + azoxystrobin (Union) 0.79 SC	5.75	28
	fluoxastrobin (Fame)		
	4 SC	0.28 to 0.36	21 to 28
	0.25 G	2.3 to 4.6 lbs	28
	fluoxastrobin + myclobutanil (Fame M) 3.9 SC	0.5 to 1	21 to 28
	fluoxastrobin + tebuconazole (Fame T) 4 SC*	0.45 to 0.9	21 to 28
	flutolanil		
	(Prostar) 70 WP, 70 WDG	2.2 to 4.5	21 to 30
	(Pedigree) 3.8 SC	3.25 to 6.6	21 to 30
	flutolanil + thiophanate-methyl (SysStar) 80 WDG	3 to 6.12	21 to 28
	isofetamid + tebuconazole (Tekken) 1.8 SC*	3	14 to 28
	mandestrobin (Pinpoint) 4SC	0.31	14
	mefentrifluconazole (Maxtima) 3.34 SC*	0.8	28
	mefentrifluconazole + pyraclostrobin (Navicon) 3.34 SC*	0.85	28
	metconazole (Tourney) 50 WDG	0.37	21
	penthiopyrad (Velista) 50 WG	0.5 to 0.7	14 to 28
	polyoxin D		
	(Affirm) 11.3 WDG	1	7
	(Endorse) 2.5 WP	4	7
	prothioconazole (Densicor) 4 SC*	0.196	14 to 21
	pydiflumetofen (Posterity) 1.67 SC*	0.08 to 0.32	21 to 28
	pydiflumetofen + azoxystrobin + propiconazole (Posterity XT) 1.48 SE*	1.5 to 3	14 to 28

41

Turfgrass Disease Control (continued)

Disease	Fungicide and Formulation[1]	Amount of Formulation (oz/1,000 sq ft)[2]	Application Interval (days)[3]
Fairy Ring	pyraclostrobin (Insignia)		
(*Basidiomycetes*)	20 WG	0.9	28
(continued)	2 SC	0.7	28
	pyraclostrobin + boscalid (Honor) 28 WG*	1.1	28
	pyraclostrobin + fluxapyroxad (Lexicon Intrinsic) 4.17 SC	0.47	28
	pyraclostrobin + triticonazole (Pillar) 0.81 G	3 lbs	14 to 28
	tebuconazole*		
	(Torque) 3.6 F	0.6 to 1.1	21
	(Mirage Stressgard) 2 SC	1 to 2	28
	triadimefon (Bayleton) 50DF, 4.15 F	1 to 2	14 to 21
Gray Leaf Spot	azoxystrobin		
(*Pyricularia grisea*)	(Heritage, Strobe) 50 WG	0.2 to 0.4	14 to 28
	(Heritage) 0.8 TL	1 to 2	14 to 28
	(Heritage) 0.31 G	2 to 4 lbs	14 to 28
	(Strobe) 2 L	0.38 to 0.77	14 to 28
	azoxystrobin + acibenzolar-S-methyl (Heritage Action) 51 WG*	0.2 to 0.4	14 to 28
	azoxystrobin + chlorothalonil (Renown) 5.16 SC*	2.5 to 4.5	10 to 14
	azoxystrobin + difenoconazole (Briskway) 2.7 SC*	0.5 to 1.2	14 to 21
	azoxystrobin + propiconazole (Headway)		
	1.4 ME	1.5 to 3	14 to 28
	1.06 G	2 to 4 lbs	14 to 28
	azoxystrobin + tebuconazole (Strobe T) 2.67 SC*	0.75	1.5
	benzovindiflupyr + difenoconazole (Ascernity) 0.86 SL*	1	14
	boscalid + chlorothalonil (Encartis) 6.25 SC*	3 to 4	14
	chlorothalonil*		
	(Daconil Ultrex) 82.5 WDG	1.8 to 3.25	7 to 21
		3.7 to 5	14
	(Daconil Weather Stik, Legend) 6 F	2 to 3.6	7 to 10
		4 to 5.5	14
	(Daconil Zn) 4.16 F	3 to 5	7 to 14
		6 to 8	14
	(Chlorothalonil 500ZN) 4.17 F	3 to 5	7 to 10
		7.9	14
	(Chlorothalonil 720SFT) 6 F	2.12 to 3.5	7 to 14
		5.5	14
	(Chlorothalonil, Chlorostar) 82.5 DF	1.8 to 3.2	7 to 10
	(Pegasus) 6 L	2 to 3.6	7 to 14
	(Pegasus) 82.5 DF	1.82 to 3.25	7 to 14
	(Pegasus HPX) 6 F	2 to 3.6	7 to 14
	chlorothalonil + acibenzolar-S-methyl (Daconil Action) 6.1 F*	2 to 3.5	7 to 10
		4 to 5.4	14
	chlorothalonil + fluoxastrobin (Fame C) 4.25 SC*	3 to 5.9	14 to 28
	chlorothalonil + iprodione + thiophanate-methyl + tebuconazole	3 to 4	14 to 21
	(Enclave) 5.3 F*	7 to 8	28
	chlorothalonil + propiconazole (Concert) 4.3 SC*	3 to 5.5	7 to 14
		5.5 to 8.5	14 to 21
	chlorothalonil + propiconazole + fludioxonil (Instrata) 3.6 SC*	2.75 to 6	10 to 14
	chlorothalonil + thiophanate-methyl*		
	(Consyst) 67 WDG	2 to 8	7 to 14
	(Peregrine) 67 WDG	2 to 8	14
	(Spectro) 90 WDG	3.72 to 5.76	14
	(TM/C) 67 WDG	2 to 8	14 to 21
	cyazofamid + azoxystrobin (Union) 0.79 SC	2.9 to 5.75	14 to 28
	fluazinam + tebuconazole (Traction) 3.24 SC*	1.3	21
	fludioxonil (Medallion) 50 WP	0.25 to 0.5	14

Turfgrass Disease Control (continued)

Disease	Fungicide and Formulation[1]	Amount of Formulation (oz/1,000 sq ft)[2]	Application Interval (days)[3]
Gray Leaf Spot	fluopyram + trifloxystrobin (Exteris Stressgard) 0.27 SC	2.135 to 6	14 to 28
(*Pyricularia grisea*)	fluoxastrobin (Fame)		
(continued)	4 SC	0.18 to 0.36	14 to 28
	0.25 G	2.3 to 4.6 lbs	14 to 28
	fluoxastrobin + myclobutanil (Fame M) 3.9 SC	0.25 to 1	14 to 28
	fluoxastrobin + tebuconazole (Fame T) 4 SC*	0.45 to 0.9	21 to 28
	flutolanil + thiophanate-methyl (SysStar) 80 WDG	2 to 3	14
	flutriafol (Rayora) 1.04 L*	0.7 to 1.4	14 to 21
	isofetamid + tebuconazole (Tekken) 1.8 SC*	3	14 to 28
	mancozeb*		
	(Fore) 80 WP	8	14
	(Dithane) 75 DF	6.4 to 12.8	7 to 14
	(Pentathlon) 4 LF	9 to 14	5
	(Pentathlon) 75 DF	8	7
	(Wingman) 75 WP	8	7
	mefentrifluconazole + pyraclostrobin (Navicon) 3.34 SC*	0.7 to 0.85	14 to 28
	metconazole (Tourney) 50 WDG	0.37	14
	mineral oil (Civitas) + proprietary pigment (Civitas Harmonizer)*	(8 to 32) + (1 to 4)	7 to 21
	myclobutanil (Eagle, Siskin) 20 EW	1.2 to 2.4	14
	polyoxin D (Affirm) 11.3 WDG	0.88	7 to 14
	propiconazole (Banner MAXX, Kestrel, Propiconazole, Savvi, Strider) 1 ME	1 to 2	14
	prothioconazole (Densicor) 4 SC*	0.196	14 to 21
	pydiflumetofen + azoxystrobin + propiconazole (Posterity XT) 1.48 SE*	1.5 to 3	14 to 28
	pyraclostrobin (Insignia)		
	20 WG	0.5 to 0.9	14 to 28
	2 SC	0.4 to 0.7	14 to 28
	pyraclostrobin + boscalid (Honor) 28 WG*	0.55 to 1.1	14 to 28
	pyraclostrobin + fluxapyroxad (Lexicon Intrinsic) 4.17 SC	0.34 to 0.47	14 to 28
	pyraclostrobin + triticonazole (Pillar) 0.81 G	3 lbs	14 to 28
	tebuconazole*		
	(Torque) 3.6 F	0.6 to 1.1	21
	(Mirage Stressgard) 2 SC	1 to 2	14 to 28
	(Skylark, Tebuconazole) 3.6 F	0.6	28
	thiophanate-methyl		
	(3336) 50 WP or 4 F	4 to 6	14
	(3336 Plus) 2 F	4 to 8	14 to 28
	(3336) 2 G	6 to 9 lbs	14
	(SysTec 1998, T-Bird, TM) 85 WDG	2.35 to 3.53	14
	(SysTec 1998, T-Bird, TM) 4.5 L	3.5 to 5	14
	triadimefon (Bayleton) 50 WSP, 4.15 F	0.5 to 1	14
	trifloxystrobin (Compass) 50 WDG	0.15 to 0.2	14
		0.25	21
	trifloxystrobin + triadimefon		
	(Armada) 50 WP	0.6 to 1.2	14 to 28
	(Tartan) 2 SC*	1 to 2	14 to 28
Helminthosporium Leaf Spot/	azoxystrobin		
Melting Out (*Bipolaris spp.;*	(Heritage, Strobe) 50 WG	0.2 to 0.4	14 to 21
Drechslera spp.)	(Heritage) 0.8 TL	1 to 2	14 to 21
	(Heritage) 0.31 G	2 to 4 lbs	14 to 21
	(Strobe) 2 L	0.38 to 0.77	14 to 21
	azoxystrobin + acibenzolar-S-methyl (Heritage Action) 51 WG*	0.2 to 0.4	14 to 21
	azoxystrobin + chlorothalonil (Renown) 5.16 SC*	2.5 to 4.5	14 to 21
	azoxystrobin + difenoconazole (Briskway) 2.7 SC*	0.5 to 1.2	14 to 21

Turfgrass Disease Control (continued)

Disease	Fungicide and Formulation[1]	Amount of Formulation (oz/1,000 sq ft)[2]	Application Interval (days)[3]
Helminthosporium Leaf Spot/ Melting Out (*Bipolaris spp.; Drechslera spp.*) (continued)	azoxystrobin + propiconazole (Headway)		
	1.4 ME	1.5 to 3	14 to 21
	1.06 G	2 to 4 lbs	14 to 21
	azoxystrobin + tebuconazole (Strobe T) 2.67 SC*	0.75 to 1.5	14 to 21
	benzovindiflupyr + difenoconazole (Ascernity) 0.86 SL*	1	14
	boscalid + chlorothalonil (Encartis) 6.25 SC*	3 to 4	14 to 21
	chlorothalonil*		
	(Daconil Ultrex) 82.5 WDG	1.8 to 3.25	7 to 21
		3.7 to 5	14 to 21
	(Daconil Weather Stik, Legend) 6 F	2 to 3.6	7 to 21
		4 to 5.5	14
	(Daconil Zn) 4.16 F	3 to 5	7 to 21
		6 to 8	14
	(Chlorothalonil 500ZN) 4.17 F	3 to 5	7 to 10
		7.9	14
	(Chlorothalonil 720SFT) 6 F	2.12 to 3.5	7 to 10
		5.5	14
	(Chlorothalonil, Chlorostar) 82.5 DF	1.8 to 3.2	7 to 10
	(Pegasus) 6 L	2 to 3.6	7 to 14
	(Pegasus) 82.5 DF	1.82 to 3.25	7 to 14
	(Pegasus HPX) 6 F	2 to 3.6	7 to 14
	chlorothalonil + acibenzolar-S-methyl (Daconil Action) 6.1 F*	2 to 3.5	7 to 21
		4 to 5.4	14
	chlorothalonil + fluoxastrobin (Fame C) 4.25 SC*	3 to 5.9	14 to 21
	chlorothalonil + propiconazole (Concert) 4.3 SC*	3 to 5.5	7 to 14
		5.5 to 8.5	14 to 21
	chlorothalonil + propiconazole + fludioxonil (Instrata) 3.6 SC*	2.75 to 6	10 to 21
	chlorothalonil + thiophanate-methyl*		
	(Consyst) 67 WDG	2 to 8	7 to 21
	(Peregrine) 67 WDG	2 to 8	14
	(Spectro) 90 WDG	3.72 to 5.76	14
	(TM/C) 67 WDG	2 to 8	14 to 21
	cyazofamid + azoxystrobin (Union) 0.79 SC	2.9 to 5.75	14 to 21
	fluazinam (Secure) 4.17 SC*	0.5	14
	fluazinam + acibenzolar-S-methyl (Secure Action) 4.18 SC*	0.5	14
	fluazinam + tebuconazole (Traction) 3.24 SC*	1.3	14
	fludioxonil (Medallion) 50 WP	0.25 to 0.5	14 to 21
	fluoxastrobin (Fame)		
	4 SC	0.18 to 0.36	14 to 21
	0.25 G	2.3 to 4.6 lbs	14 to 21
	fluoxastrobin + myclobutanil (Fame M) 3.9 SC	0.25 to 1	14 to 28
	fluoxastrobin + tebuconazole (Fame T) 4 SC*	0.45 to 0.9	21 to 28
	flutolanil + thiophanate-methyl (SysStar) 80 WDG	2 to 3	14
	iprodione (26GT, Iprodione Pro, IPro, Raven) 2 F, 2 SC, 2 SE*	3 to 4	14 to 28
	iprodione + thiophanate-methyl *		
	(26/36) 3.8 F	2 to 4	14 to 21
	(Dovetail) 3.8 F	1 to 4	14 to 21
	iprodione + trifloxystrobin (Interface) 2.27 SC*	3 to 5	refer to label
	mancozeb*		
	(Fore) 80 WP	4	7 to 14
	(Dithane) 75 DF	4	10
	(Pentathlon) 4 LF	5 to 14	3 to 5
	(Pentathlon) 75 DF	4	7
	(Protect, Wingman) 75 WP	4	7 to 14
	mancozeb + copper hydroxide (Junction) 60 DF*	2 to 4	7 to 14

Turfgrass Disease Control (continued)

Disease	Fungicide and Formulation[1]	Amount of Formulation (oz/1,000 sq ft)[2]	Application Interval (days)[3]
Helminthosporium Leaf Spot/ Melting Out (*Bipolaris spp.; Drechslera spp.*) (continued)	mefentrifluconazole + pyraclostrobin (Navicon) 3.34 SC*	0.7 to 0.85	14 to 28
	mineral oil (Civitas) + proprietary pigment (Civitas Harmonizer)*	(8 to 32) + (1 to 4)	7 to 21
	myclobutanil (Eagle, Myclobutanil, Siskin) 20 EW	1.2	14
	penthiopyrad (Velista) 50 WG	0.3 to 0.5	14
	polyoxin D		
	(Affirm) 11.3 WDG	0.88	7 to 14
	(Endorse) 2.5 WP	4	7 to 14
	propiconazole		
	(Banner MAXX, Kestrel, Propiconazole, Savvi, Strider) 1 ME	1 to 2	14
	pydiflumetofen + azoxystrobin + propiconazole (Posterity Forte) 2.5 SE*	0.63 to 0.84	14 to 21
	pydiflumetofen + azoxystrobin + propiconazole (Posterity XT) 1.48 SE*	1.5 to 3	14 to 21
	pyraclostrobin (Insignia)		
	20 WG	0.5 to 0.9	14 to 28
	2 SC	0.4 to 0.7	14 to 28
	pyraclostrobin + boscalid (Honor) 28WG*	0.55 to 1.1	14 to 28
	pyraclostrobin + fluxapyroxad (Lexicon Intrinsic) 4.17 SC	0.34 to 0.47	14 to 28
	pyraclostrobin + triticonazole (Pillar) 0.81 G	3 lbs	14 to 28
	thiophanate-methyl		
	(3336) 50 WP or 4 F	4 to 6	14
	(3336 Plus) 2 F	4 to 8	14 to 28
	(3336) 2 G	6 to 9 lbs	14
	thiram (Spotrete) 4 F*	3.75 to 7.5	3 to 10
	trifloxystrobin (Compass) 50 WDG	0.1 to 0.15	14
		0.15 to 0.25	21 to 28
	trifloxystrobin + triadimefon		
	(Armada) 50 WP	0.6 to 1.2	14 to 28
	(Tartan) 2 SC*	1 to 2	14 to 28
	triticonazole		
	(Trinity) 1.7 SC	0.5 to 2	14 to 28
	(Triton) 70 WDG	0.15 to 0.3	14 to 28
	triticonazole + chlorothalonil (Reserve) 4.79 SC*	3.2 to 4.5	14 to 28
	vinclozolin (Curalan, Touche) 50 EG*	1	14 to 28
Large Patch (*Rhizoctonia solani*)	azoxystrobin		
	(Heritage, Strobe) 50 WG	0..2 to 0.4	14 to 28
	(Heritage) 0.8 TL	2	14 to 28
	(Heritage) 0.31 G	2 to 4 lbs	14 to 28
	(Strobe) 2 L	0.38 to 0.77	28
	azoxystrobin + acibenzolar-S-methyl (Heritage Action) 51 WG*	0.2 to 0.4	14 to 28
	azoxystrobin + chlorothalonil (Renown) 5.16 SC*	2.5	14
		4.5	14 to 21
	azoxystrobin + difenoconazole (Briskway) 2.7 SC*	0.5 to 1.2	14 to 28
	azoxystrobin + propiconazole (Headway)		
	1.4 ME	1.5 to 3	14 to 28
	1.06 G	2 to 4 lbs	14 to 28
	azoxystrobin + tebuconazole (Strobe T) 2.67 SC*	0.75 to 1.5	14 to 21
	benzovindiflupyr + difenoconazole (Ascernity) 0.86 SL*	1	14 to 21
	chloroneb*		
	(Teremec) 65 SP	5	21 to 28
	(Teremec) 2.9 F	9	21 to 28
	chlorothalonil + fluoxastrobin (Fame C) 4.25 SC*	3 to 5.9	14 to 28
	chlorothalonil + iprodione + thiophanate-methyl + tebuconazole	3 to 4	14 to 21
	(Enclave) 5.3 F*	7 to 8	28
	chlorothalonil + thiophanate-methyl*		
	(Consyst) 67 WDG	2 to 8	7 to 14
	(Peregrine) 67 WDG	2 to 8	14

Turfgrass Disease Control (continued)

Disease	Fungicide and Formulation[1]	Amount of Formulation (oz/1,000 sq ft)[2]	Application Interval (days)[3]
Large Patch	cyazofamid + azoxystrobin (Union) 0.79 SC	5.75	14 to 28
(*Rhizoctonia solani*)	fluazinam (Secure) 4.17 SC*	0.5	14
(continued)	fluazinam + acibenzolar-S-methyl (Secure Action) 4.18 SC*	0.5	14
	fluazinam + tebuconazole (Traction) 3.24 SC*	1.3	14
	fluoxastrobin (Fame)		
	4 SC	0.28 to 0.36	14 to 28
	0.25	2.3 to 4.6 lbs	14 to 28
	fluoxastrobin + myclobutanil (Fame M) 3.9 SC	0.5 to 1	21 to 28
	fluxapyroxad (Xzemplar) 2.47 SC	0.21 to 0.26	14 to 28
	flutolanil		
	(Prostar) 70 WP, 70 WDG	2.2	30
	(Pedigree) 3.8 SC	3.25	30
	flutriafol (Rayora) 1.04 L*	0.7 to 1.4	28
	iprodione (26GT, Iprodione Pro, IPro, Raven) 2 F, 2 SC, 2 SE*	4	14 to 21
	iprodione + thiophanate-methyl (26/36) 3.8 F*	2 to 4	14 to 21
	iprodione + trifloxystrobin (Interface) 2.27 SC*	4	14 to 21
	isofetamid + tebuconazole (Tekken) 1.8 SC*	3	14 to 28
	mefentrifluconazole + pyraclostrobin (Navicon) 3.34 SC*	0.7 to 0.85	14 to 28
	metconazole (Tourney) 50 WDG	0.37	14
	myclobutanil (Eagle, Myclobutanil, Siskin) 20 EW	2.4	28 (fall)
	penthiopyrad (Velista) 50 WG	0.7	14 to 28
	polyoxin D		
	(Affirm) 11.3 WDG	0.88	7 to 14
	(Endorse) 2.5 WP	4	7 to 14
	propiconazole (Banner MAXX, Kestrel, Propiconazole, Savvi, Strider) 1 ME	3 to 4	early fall
	prothioconazole (Densicor) 4 SC*	0.196	14 to 28
	pydiflumetofen + azoxystrobin + propiconazole (Posterity Forte) 2.5 SE*	0.84	14 to 21
	pydiflumetofen + azoxystrobin + propiconazole (Posterity XT) 1.48 SE*	1.5 to 3	14 to 28
	pyraclostrobin (Insignia)		
	20 WG	0.5 to 0.9	14 to 28
	2 SC	0.4 to 0.7	14 to 28
	pyraclostrobin + boscalid (Honor) 28 WG*	1.1	14 to 28
	pyraclostrobin + fluxapyroxad (Lexicon Intrinsic) 4.17 SC	0.34 to 0.47	14 to 28
	pyraclostrobin + triticonazole (Pillar) 0.81 G	3 lbs	14 to 28
	tebuconazole*		
	(Torque) 3.6 F	0.6 to 1.1	21
	(Mirage Stressgard) 2 SC	1 to 2	28
	(Skylark, Tebuconazole) 3.6 F	0.6	28
	thiophanate-methyl		
	(3336) 50WP or 4 F	2 to 4	14
	(3336 Plus) 2 F	2 to 4	14 to 28
	(SysTec 1998, T-Bird, TM) 85 WDG	0.67 to 1.3	14
	(3336) 2 G	1.5 to 6 lbs	14
	(SysTec 1998, T-Bird, TM) 4.5 L	1 to 2	14
	thiophanate-methyl + flutolanil (SysStar) 80 WDG	2 to 3	14 to 21
	triadimefon (Bayleton) 50 WSP, 4.15 F	1 to 2	fall and spring
	triticonazole		
	(Trinity) 1.7 SC	1 to 2	14 to 28
	(Triton) 70 WDG	0.15 to 0.3	14 to 28
	triticonazole + chlorothalonil (Reserve) 4.79 SC*	3.2 to 5.4	14 to 28

Turfgrass Disease Control (continued)

Disease	Fungicide and Formulation[1]	Amount of Formulation (oz/1,000 sq ft)[2]	Application Interval (days)[3]
Leaf and Sheath Spot (*Rhizoctonia zeae, R. oryzae*)	azoxystrobin (Heritage)		
	0.8 TL	2	14 to 28
	0.31 G	2 to 4 lbs	14 to 28
	azoxystrobin + acibenzolar-S-methyl (Heritage Action) 51 WG*	0.2 to 0.4	14 to 28
	azoxystrobin + chlorothalonil (Renown) 5.16 SC*	2.5	14
		4.5	14 to 21
	azoxystrobin + difenoconazole (Briskway) 2.7 SC*	0.5 to 1.2	14 to 28
	azoxystrobin + propiconazole (Headway)		
	1.4 ME	1.5 to 3	14 to 28
	1.06 G	2 to 4 lbs	14 to 28
	azoxystrobin + tebuconazole (Strobe T) 2.67 SC*	0.75 to 1.5	14 to 21
	benzovindiflupyr + difenoconazole (Ascernity) 0.86 SL*	1	14 to 21
	chlorothalonil + propiconazole + fludioxonil (Instrata) 3.59 SC*	2.75 to 6	14 to 21
	chlorothalonil + thiophanate-methyl (Spectro) 90 WDG*	3 to 5.76	14 to 21
	flutolanil		
	(Prostar) 70 WDG	2.2 to 4.5	14 to 21
	(Pedigree) 3.8 SC	3.25 to 6.6	14 to 21
	mefentrifluconazole + pyraclostrobin (Navicon) 3.34 SC*	0.7 to 0.85	14 to 28
	penthiopyrad (Velista) 50 WG	0.3 to 0.5	14
	polyoxin D (Affirm) 11.3 WDG	0.88	7 to 14
	prothioconazole (Densicor) 4 SC*	0.196	14 to 21
	pydiflumetofen + azoxystrobin + propiconazole (Posterity XT) 1.48 SE*	1.5 to 3	14 to 28
	pyraclostrobin (Insignia)		
	20 WG	0.5 to 0.9	14 to 28
	2 SC	0.4 to 0.7	14 to 28
	pyraclostrobin + boscalid (Honor) 28 WG*	1.1	14 to 28
	pyraclostrobin + fluxapyroxad (Lexicon Intrinsic) 4.17 SC	0.34 to 0.47	14 to 28
	pyraclostrobin + triticonazole (Pillar) 0.81 G	3 lbs	28
Pink Patch (*Limonomyces roseipelis*)	azoxystrobin		
	(Heritage, Strobe) 50 WG	0.2 to 0.4	14 to 28
	(Heritage) 0.8 TL	1 to 2	14 to 28
	(Heritage) 0.31 G	2 to 4 lbs	14 to 28
	(Strobe) 2 L	0.38 to 0.77	14 to 28
	azoxystrobin + acibenzolar-S-methyl (Heritage Action) 51 WG*	0.2 to 0.4	14 to 28
	azoxystrobin + chlorothalonil (Renown) 5.16 SC*	2.5 to 4.5	14 to 21
	azoxystrobin + difenoconazole (Briskway) 2.7 SC*	0.5 to 1.2	14 to 28
	azoxystrobin + propiconazole (Headway)		
	1.4 ME	1.5 to 3	14 to 28
	1.06 G	2 to 4 lbs	14 to 28
	azoxystrobin + tebuconazole (Strobe T) 2.67 SC*	0.75 to 1.5	14 to 21
	benzovindiflupyr + difenoconazole (Ascernity) 0.86 SL*	1	14
	chlorothalonil + fluoxastrobin (Fame C) 4.25 SC*	3 to 5.9	14 to 28
	chlorothalonil + propiconazole (Concert) 4.3 SC*	3 to 5.5	7 to 14
		5.5 to 8.5	14 to 21
	cyazofamid + azoxystrobin (Union) 0.79 SC	2.9 to 5.75	14 to 28
	fluazinam (Secure) 4.17 SC*	0.5	14
	fluazinam + acibenzolar-S-methyl (Secure Action) 4.18 SC*	0.5	14
	fluazinam + tebuconazole (Traction) 3.24 SC*	1.3	14
	fluopyram + trifloxystrobin (Exteris Stressgard) 0.27 SC	1.5 to 4.135	14 to 28
	fluoxastrobin (Fame)		
	4 SC	0.18 to 0.36	14 to 28
	0.25 G	2.3 to 4.6 lbs	14 to 28
	flutolanil		
	(Prostar) 70 WP, 70 DG	1.5	21 to 28
	(Pedigree) 3.8 SC	2.2	21 to 28
	flutolanil + thiophanate-methyl (SysStar) 80 WDG	2	21 to 28

Turfgrass Disease Control (continued)

Disease	Fungicide and Formulation[1]	Amount of Formulation (oz/1,000 sq ft)[2]	Application Interval (days)[3]
Pink Patch	iprodione + trifloxystrobin (Interface) 2.27 SC*	3 to 4	14
(*Limonomyces roseipelis*)	isofetamid + tebuconazole (Tekken) 1.8 SC*	3	14 to 28
(continued)	mefentrifluconazole + pyraclostrobin (Navicon) 3.34 SC*	0.7 to 0.85	14 to 28
	propiconazole (Banner MAXX, Kestrel, Propiconazole, Savvi, Strider) 1 ME	2	14 to 21
	pydiflumetofen + azoxystrobin + propiconazole (Posterity XT) 1.48 SE*	1.5 to 3	14 to 28
	pyraclostrobin (Insignia)		
	20 WG	0.5 to 0.9	14 to 28
	2 SC	0.4 to 0.7	14 to 28
	pyraclostrobin + boscalid (Honor) 28 WG*	0.55 to 1.1	14 to 28
	pyraclostrobin + fluxapyroxad (Lexicon Intrinsic) 4.17 SC	0.34 to 0.47	14 to 28
	pyraclostrobin + triticonazole (Pillar) 0.81 G	3 lbs	14 to 28
	tebuconazole*		
	(Torque) 3.6 F	0.6 to 1.1	refer to label
	(Mirage Stressgard) 2 SC	1 to 2	14 to 28
	(Skylark, Tebuconazole) 3.6 F	0.6	28
	trifloxystrobin (Compass) 50 WDG	0.1 to 0.15	14
		0.2 to 0.25	21
	trifloxystrobin + triadimefon		
	(Armada) 50 WP	0.6 to 1.2	14 to 28
	(Tartan) 2 SC*	1 to 2	14 to 28
	triticonazole (Trinity) 1.7 EC	1 to 2	14 to 28
	triticonazole + chlorothalonil (Reserve) 4.79 SC*	3.2 to 4.5	refer to label
	vinclozolin (Curalan, Touche) 50 EG*	1	14 to 28
Pink Snow Mold/Microdochium Patch (*Microdochium nivale*)	azoxystrobin		
	(Heritage, Strobe) 50 WG	0.2 to 0.4	10 to 28
		0.7	1 application
	(Heritage) 0.8 TL	2	10-28
		3.5	1 application
	(Heritage) 0.31 G	4 lbs	10 to 28
		7 lbs	1 application
	(Strobe) 2L	0.77	14
		1.35	1 application
	azoxystrobin + acibenzolar-S-methyl (Heritage Action) 51 WG*	0.4	refer to label
	azoxystrobin + chlorothalonil (Renown) 5.16 SC*	2.5 to 4.5	14 to 21
	azoxystrobin + difenoconazole (Briskway) 2.7 SC*	0.5 to 1.2	14 to 28
	azoxystrobin + propiconazole (Headway)		
	1.4 ME	1.5 to 3	10 to 28
		5.25	1 application
	1.06 G	2 to 4 lbs	14 to 28
		5	1 application
	azoxystrobin + tebuconazole (Strobe T) 2.67 SC*	0.75 to 1.5	14 to 21
		2.4	1 application
	benzovindiflupyr + difenoconazole (Ascernity) 0.86 SL*	1	14
	chlorothalonil + acibenzolar-S-methyl (Daconil Action) 6.1 F*	5.4	21 to 28
	chlorothalonil + fluoxastrobin (Fame C) 4.25 SC*	3 to 5.9	28
	chlorothalonil + iprodione + thiophanate-methyl + tebuconazole (Enclave) 5.3 F*	7 to 8	28
	chlorothalonil + propiconazole (Concert) 4.3 SC*	8.5	14 to 28
	chlorothalonil + propiconazole + fludioxonil (Instrata) 3.6 SC*	5 to 11	late fall
	chlorothalonil + thiophanate-methyl*		
	(Consyst, Peregrine, TM/C) 67 WDG	6 to 8	1 application
	(Spectro) 90 WDG	3.72 to 5.76	14
	fenarimol (Rubigan) 1 AS*	8	1 application
		4	30 (2 applications)

Turfgrass Disease Control (continued)

Disease	Fungicide and Formulation[1]	Amount of Formulation (oz/1,000 sq ft)[2]	Application Interval (days)[3]
Pink Snow Mold/Microdochium Patch (*Microdochium nivale*) (continued)	fluazinam (Secure) 4.17 SC*	0.5	late fall
	fluazinam + acibenzolar-S-methyl (Secure Action) 4.18 SC*	0.5	late fall
	fluazinam + tebuconazole (Traction) 3.24 SC*	1.3	late fall
	fludioxonil (Medallion) 50 WP	0.25 to 0.5	14
	fluopyram + trifloxystrobin (Exteris Stressgard) 0.27 SC	4.135 to 12.6	10 to 28
	fluoxastrobin (Fame)		
	4 SC	0.18 to 0.36	14 to 28
	0.25 G	2.3 to 4.6 lbs	14 to 28
	fluoxastrobin + myclobutanil (Fame M) 3.9 SC	0.5 to 1	21 to 28
	fluoxastrobin + tebuconazole (Fame T) 4 SC*	0.45 to 0.9	30
	fluxapyroxad (Xzemplar) 2.47 SC	0.26	14 to 28
	flutolanil + thiophanate-methyl (SysStar) 80 WDG	4 to 6.12	1 application
		2 to 3	14 to 21
	iprodione (26GT, Iprodione Pro, IPro, Raven) 2 F, 2 SC, 2 SE*	4 to 8	1 to 2 applications
	iprodione + thiophanate-methyl*		
	(26/36) 3.8 F	2 to 4	14 to 21
	(Dovetail) 3.8 F	1 to 4	14 to 21
	iprodione + trifloxystrobin (Interface) 2.27 SC*	4 to 7	1 application
	isofetamid + tebuconazole (Tekken) 1.8 SC*	3	14 to 28
	mancozeb*		
	(Fore) 80 WP	6 to 8	14 to 42
	(Dithane, Pentathlon) 75 DF	6 to 8	14 to 42
	(Pentathlon) 4 LF	10 to 14	14 to 42
	(Protect) 75 WP	6 to 8	7 to 14
	mancozeb + copper hydroxide (Junction) 60 DF*	2 to 4	14 to 42
	mefentrifluconazole + pyraclostrobin (Navicon) 3.34 SC*	0.7 to 0.85	14 to 28
	metconazole (Tourney) 50 WDG	0.37 to 0.44	late fall
	mineral oil (Civitas) + proprietary pigment (Civitas Harmonizer)*	(8 to 32) + (1 to 4)	7 to 21
	myclobutanil (Eagle, Myclobutanil, Siskin) 20 EW	1.2 to 2.4	prior to snow cover
	PCNB (various brands)		
	75 WP	3 to 8	28 to 42
	10 G	80 to 160	prior to snowfall
	4 F	12 to 16	prior to snowfall
	polyoxin D		
	(Affirm) 11.3 WDG	0.88	7 to 14
	(Endorse) 2.5 WP	4	7 to 14
	propiconazole (Banner MAXX, Kestrel, Propiconazole, Savvi, Strider) 1 ME	2 to 4	fall to early spring
	prothioconazole (Densicor) 4 SC*	0.196	14 to 21
	pydiflumetofen (Posterity) 1.67 SC*	0.08 to 0.16	14 to 28
	pydiflumetofen + azoxystrobin + propiconazole (Posterity XT) 1.48 SE*	1.5 to 3	14 to 28
	pyraclostrobin (Insignia)		
	20 WG	0.5 to 0.9	14 to 28
	2 SC	0.7	14 to 28
	pyraclostrobin + boscalid (Honor) 28 WG*	0.55 to 1.1	14 to 28
	pyraclostrobin + fluxapyroxad (Lexicon Intrinsic) 4.17 SC	0.47	14 to 28
	pyraclostrobin + triticonazole (Pillar) 0.81 G	3 lbs	28
	tebuconazole*		
	(Torque) 3.6 F	0.6 to 1.1	prior to snowfall
	(Mirage Stressgard) 2 SC	1 to 2	10 to 28
	(Skylark, Tebuconazole) 3.6 F	0.6	prior to snowfall
	thiram (Spotrete)*		
	4 F	3 to 12	fall and spring
	75 WDG	3 to 8	fall and spring

Turfgrass Disease Control (continued)

Disease	Fungicide and Formulation[1]	Amount of Formulation (oz/1,000 sq ft)[2]	Application Interval (days)[3]
Pink Snow Mold/Microdochium Patch (*Microdochium nivale*) (continued)	thiophanate-methyl		
	(3336) 50WP or 4 F	2 to 4	14
	(3336 Plus) 2 F	2 to 4	14 to 28
	(SysTec 1998, T-Bird, TM) 85 WDG	0.67 to 1.3	14
	(3336) 2 G	1.5 to 6 lbs	14
	(SysTec 1998, T-Bird, TM) 4.5 L	1 to 2	14
	triadimefon (Bayleton) 50 WSP, 4.15 F	1 to 2	60 to 90
	trifloxystrobin (Compass) 50 WDG	0.2 to 0.25	fall to early spring
	trifloxystrobin + triadimefon		
	(Armada) 50 WP	1.2	fall to early spring
	(Tartan) 2 SC*	2	fall to early spring
	triticonazole		
	(Trinity) 1.7 SC	0.5 to 2	14 to 28
	(Triton) 70 WDG	0.15 to 0.3	late fall
	(Triton Flo) 3 G	0.28 to 1.1	10 to 14
	triticonazole + chlorothalonil (Reserve) 4.79 SC*	3.2 to 4.5	14 to 28
	vinclozolin (Curalan, Touche) 50 EG*	1	10 to 21
Powdery Mildew (*Blumeria graminis*)	azoxystrobin		
	(Heritage, Strobe) 50 WG	0.2 to 0.4	14 to 28
	(Heritage) 0.8 TL	1 to 2	14 to 28
	(Heritage) 0.31 G	2 to 4 lbs	14 to 28
	azoxystrobin + acibenzolar-S-methyl (Heritage Action) 51 WG*	0.2 to 0.4	14 to 28
	azoxystrobin + chlorothalonil (Renown) 5.16 SC*	2.5 to 4.5	14 to 21
	azoxystrobin + difenoconazole (Briskway) 2.7 SC*	0.5 to 1.2	14 to 28
	azoxystrobin + propiconazole (Headway)		
	1.4 ME	1.5 to 3	14 to 28
	1.06 G	2 to 4 lbs	14 to 28
	azoxystrobin + tebuconazole (Strobe T) 2.67 SC*	0.75 to 1.5	14 to 21
	benzovindiflupyr + difenoconazole (Ascernity) 0.86 SL*	1	14 to 21
	chlorothalonil + fluoxastrobin (Fame C) 4.25 SC*	3 to 5.9	14 to 28
	chlorothalonil + propiconazole (Concert) 4.3 SC*	4.5 to 8.5	14 to 28
	chlorothalonil + thiophanate-methyl*		
	(Consyst) 67 WDG	2 to 8	7 to 14
	(Peregrine) 67 WDG	2 to 8	14
	(Spectro) 90 WDG	3.72 to 5.76	14
	(TM/C) 67 WDG	2 to 8	14 to 21
	fenarimol (Rubigan) 1 AS*	2 to 4	1 application
	fluazinam + tebuconazole (Traction) 3.24 SC*	1.3	14
	fluoxastrobin (Fame)		
	4 SC	0.18 to 0.36	14 to 28
	0.25 G	2.3 to 4.6 lbs	14 to 28
	fluoxastrobin + myclobutanil (Fame M) 3.9 SC	0.25 to 1	14 to 28
	flutriafol (Rayora) 1.04 L*	0.7 to 1.4	14 to 21
	isofetamid + tebuconazole (Tekken) 1.8 SC*	3	14 to 28
	mancozeb + copper hydroxide (Junction) 60 DF*	2 to 4	7 to 14
	mefentrifluconazole + pyraclostrobin (Navicon) 3.34 SC*	0.7 to 0.85	14 to 28
	mineral oil (Civitas) + proprietary pigment (Civitas Harmonizer)*	(8 to 32) + (1 to 4)	7 to 21
	myclobutanil (Eagle, Myclobutanil, Siskin) 20 EW	1.2	14 to 28
	penthiopyrad (Velista) 50 WG	0.3 to 0.5	14
	propiconazole (Banner MAXX, Kestrel, Propiconazole, Savvi, Strider) 1 ME	1 to 2	14 to 28
	pydiflumetofen + azoxystrobin + propiconazole (Posterity XT) 1.48 SE*	1.5 to 3	14 to 28
	pyraclostrobin (Insignia)		
	20 WG	0.5 to 0.9	14 to 28
	2 SC	0.4 to 0.7	14 to 28

Turfgrass Disease Control (continued)

Disease	Fungicide and Formulation[1]	Amount of Formulation (oz/1,000 sq ft)[2]	Application Interval (days)[3]
Powdery Mildew (*Blumeria graminis*) (continued)	pyraclostrobin + boscalid (Honor) 28 WG*	0.55 to 1.1	14 to 28
	pyraclostrobin + fluxapyroxad (Lexicon Intrinsic) 4.17 SC	0.34 to 0.47	14 to 28
	pyraclostrobin + triticonazole (Pillar) 0.81 G	3 lbs	14 to 28
	tebuconazole*		
	(Torque) 3.6 F	0.6 to 1.1	refer to label
	(Skylark, Tebuconazole) 3.6 F	0.6	28
	triadimefon (Bayleton) 50 WSP, 4.15 F	0.5 to 1	15 to 30
Pythium Blight (*Pythium aphanidermatum*)	azoxystrobin		
	(Heritage, Strobe) 50 WG	0.4	10 to 14
	(Heritage) 0.8 TL	2	10 to 14
	(Heritage) 0.31 G	2 to 4 lbs	10 to 14
	(Strobe) 2 L	0.38 to 0.77	10 to 14
	azoxystrobin + acibenzolar-S-methyl (Heritage Action) 51 WG*	0.2 to 0.4	10 to 14
	azoxystrobin + propiconazole (Headway)		
	1.4 ME	3	10 to 14
	1.06 G	2 to 4 lbs	14 to 28
	azoxystrobin + tebuconazole (Strobe T) 2.67 SC*	0.75 to 1.5	10 to 21
	chloroneb*		
	(Teremec) 65 SP	4	5 to 7
	(Teremec) 2.9 F	7	5 to 7
	chlorothalonil + fluoxastrobin (Fame C) 4.25 SC*	3 to 5.9	7 to 14
	cyazofamid (Segway) 3.33 SC	0.45 to 0.9	14 to 21
	cyazofamid + azoxystrobin (Union) 0.79 SC	2.9 to 5.75	14 to 28
	ethazole*		
	(Koban) 30 WP	2 to 4.5	10
	(Terrazole) 35 WP	2 to 4	10 to 14
	fluopicolide + propamocarb (Stellar) 5.7 SC	1.2	14
	fluoxastrobin (Fame)		
	4 SC	0.18 to 0.36	7 to 14
	0.25 G	2.3 to 4.6 lbs	14
	fluoxastrobin + myclobutanil (Fame M) 3.9 SC	0.5 to 1	14
	fluoxastrobin + tebuconazole (Fame T) 4 SC*	0.45 to 0.9	21
	fosetyl Al		
	(Signature, Fosetyl-Al) 80 WDG	4 to 8	14 to 21
	(Signature Xtra Stressgard) 60 WDG*	2 to 6	7 to 21
	(Autograph) 70 DF*	4.6 to 9.2	14 to 21
	(Viceroy) 70 DF	4.6 to 9.1	14 to 21
	mancozeb*		
	(Fore) 80 WP	8	5 to 14
	(Dithane) 75 DF	8	10
	(Pentathlon) 4 LF	14	5
	(Pentathlon) 75 DF	8	5
	(Protect, Wingman) 75 WP	8	7 to 14
	mancozeb + copper hydroxide (Junction) 60 DF*	2 to 4	5
	mefenoxam		
	(Subdue) 43 WSP	0.28 to 0.56	10 to 21
	(Subdue MAXX, Quell) 2 ME	0.5 to 1	10 to 21
	(Subdue) 1 GR	12.5 to 25	10 to 14
	(Fenox, Mefenoxam) 2 AQ, 2 EC	0.2 to 1	10 to 21
	mefentrifluconazole + pyraclostrobin (Navicon) 3.34 SC*	0.85	10 to 14
	metalaxyl (Vireo) 2 MEC	1 to 2	10 to 21

Turfgrass Disease Control (continued)

Disease	Fungicide and Formulation[1]	Amount of Formulation (oz/1,000 sq ft)[2]	Application Interval (days)[3]
Pythium Blight	phosphorus acid		
(*Pythium aphanidermatum*)	(Alude, Resyst) 3.3 F	5 to 10	7 to 14
(continued)	(Jetphiter) 5.41 F	5	7
	(Magellan) 4.3 F	4.1 to 8.2	14 to 21
	(Vital) 4.2 F	4 to 6	14
	(Vital Sign) 4.2 F	4 to 8	7 to 14
	potassium phosphite (Appear, Appear II) 4.1 SC	3 to 4	7 to 14
		4 to 6	14
	propamocarb (Banol) 6 S*	1.3 to 4	7 to 21
	Pseudomonas chlororaphis strain AFS009 (Zio) SC	1.8 to 6.0	7 to 21
	pydiflumetofen + azoxystrobin + propiconazole (Posterity XT) 1.48 SE*	3	14
	pyraclostrobin (Insignia)		
	20 WG	0.9	14 to 28
	2 SC	0.7	10 to 14
	pyraclostrobin + boscalid (Honor) 28 WG*	1.1	10 to 14
	pyraclostrobin + fluxapyroxad (Lexicon Intrinsic) 4.17 SC	0.47	14
	pyraclostrobin + triticonazole (Pillar) 0.81 G	3 lbs	14
Pythium Root Dysfunction	azoxystrobin (Heritage)		
(*Pythium volutum*)	50 WG	0.4	21 to 28
	0.8 TL	2	21 to 28
	azoxystrobin + acibenzolar-S-methyl (Heritage Action) 51 WG*	0.4	21 to 28
	cyazofamid (Segway) 3.33 SC	0.9	14 to 28
	cyazofamid + azoxystrobin (Union) 0.79 SC	5.75	21 to 28
	fluoxastrobin (Fame)		
	4 SC	0.27 to 0.36	14 to 28
	0.25 G	3.6 to 4.6 lbs	14 to 28
	fluoxastrobin + chlorothalonil (Fame C) 4.25 SC *	4.5 to 5.9	14 to 28
	fluoxastrobin + myclobutanil (Fame M) 3.9 SC	0.5 to 1	14 to 28
	mefentrifluconazole + pyraclostrobin (Navicon) 3.34 SC*	0.85	14 to 28
	pydiflumetofen + azoxystrobin + propiconazole (Posterity XT) 1.48 SE*	3	21 to 28
	pyraclostrobin (Insignia)		
	20 WG	0.9	14 to 28
	2 SC	0.7	14 to 28
	pyraclostrobin + boscalid (Honor) 28WG*	1.1	14 to 28
	pyraclostrobin + fluxapyroxad (Lexicon Intrinsic) 4.17 SC	0.47	14 to 28
	pyraclostrobin + triticonazole (Pillar) 0.81 G	3 lbs	14
Pythium Root Rot	azoxystrobin		
(*Pythium spp.*)	(Heritage, Strobe) 50 WG	0.4	10 to 14
	(Heritage) 0.8 TL	2	10 to 14
	(Heritage) 0.31 G	2 to 4	10 to 14
	(Strobe) 2 L	0.38 to 0.77	10 to 14
	azoxystrobin + propiconazole (Headway)		
	1.4 ME	3	10 to 14
	1.06 G	2 to 4 lbs	14 to 28
	azoxystrobin + tebuconazole (Strobe T) 2.67 SC*	0.75 to 1.5	10 to 21
	chlorothalonil + fluoxastrobin (Fame C) 4.25 SC *	3 to 5.9	7 to 10
	cyazofamid (Segway) 3.33 SC	0.9	21
	cyazofamid + azoxystrobin (Union) 0.79 SC	2.9 to 5.75	14 to 21
	ethazole*		
	(Koban) 30 WP	4.5	10
	(Terrazole) 35 WP	2 to 4	10 to 14
	fluoxastrobin (Fame)		
	4 SC	0.18 to 0.36	7 to 10
	0.25 G	2.3 to 4.6 lbs	14
	fluoxastrobin + myclobutanil (Fame M) 3.9 SC	0.5 to 1	14

Turfgrass Disease Control (continued)

Disease	Fungicide and Formulation[1]	Amount of Formulation (oz/1,000 sq ft)[2]	Application Interval (days)[3]
Pythium Root Rot (*Pythium spp.*) (continued)	fluoxastrobin + tebuconazole (Fame T) 4 SC*	0.45 to 0.9	21
	fosetyl Al		
	(Signature, Fosetyl-Al) 80 WDG	4 to 8	14 to 21
	(Signature Xtra Stressgard) 60 WDG*	2 to 6	7 to 21
	(Autograph) 70 DF*	4.6 to 9.2	14 to 21
	(Viceroy) 70 DF	4.6 to 9.1	14 to 21
	mefentrifluconazole + pyraclostrobin (Navicon) 3.34 SC*	0.85	14 to 28
	phosphorous acid		
	(Vital Sign) 2.4 F	6 to 8	7 to 14
	(Jetphiter) 5.41 F	3.5 to 5	7 to 28
	potassium phosphite (Appear) 4.1 SC	6 to 8	7 to 14
	propamocarb (Banol) 6 S*	1.3 to 4	7 to 21
	Pseudomonas chlororaphis strain AFS009 (Zio) SC	1.8 to 6.0	7 to 21
	pydiflumetofen + azoxystrobin + propiconazole (Posterity XT) 1.48 SE*	3	14
Rapid Blight (*Labyrinthula spp.*)	iprodione + trifloxystrobin (Interface) 2.27 SC*	3 to 5	refer to label
	mancozeb (Fore) 80 WP*	8	14
	mefentrifluconazole + pyraclostrobin (Navicon) 3.34 SC*	0.7 to 0.85	14 to 28
	penthiopyrad (Velista) 50 WG	0.5	14
	pyraclostrobin (Insignia)		
	20 WG	0.5 to 0.9	14
	2 SC	0.4 to 0.7	14 to 28
	pyraclostrobin + boscalid (Honor) 28WG*	0.55 to 1.1	14 to 28
	pyraclostrobin + fluxapyroxad (Lexicon Intrinsic) 4.17 SC	0.34 to 0.47	14
	pyraclostrobin + triticonazole (Pillar) 0.81 G	3 lbs	14 to 28
	trifloxystrobin (Compass) 50 WDG	0.15 to 0.2	14
		0.25	21
	trifloxystrobin + triadimefon (Armada) 50 WP	0.6 to 1.2	14 to 28
Red Thread (*Laetisaria fuciformis*)	azoxystrobin		
	(Heritage, Strobe) 50 WG	0.2 to 0.4	14 to 28
	(Heritage) 0.8 TL	1 to 2	14 to 28
	(Heritage) 0.31 G	2 to 4 lbs	14 to 28
	(Strobe) 2 L	0.38 to 0.77	14 to 28
	azoxystrobin + acibenzolar-S-methyl (Heritage Action) 51 WG*	0.2 to 0.4	14 to 28
	azoxystrobin + chlorothalonil (Renown) 5.16 SC*	2.5 to 4.5	14 to 21
	azoxystrobin + difenoconazole (Briskway) 2.7 SC*	0.5 to 1.2	14 to 28
	azoxystrobin + propiconazole (Headway)		
	1.4 ME	1.5 to 3	14 to 28
	1.06 G	2 to 4 lbs	14 to 28
	azoxystrobin + tebuconazole (Strobe T) 2.67 SC*	0.75 to 1.5	14 to 21
	benzovindiflupyr + difenoconazole (Ascernity) 0.86 SL*	1	14
	boscalid + chlorothalonil (Encartis) 6.25 SC*	3 to 4	14

Turfgrass Disease Control (continued)

Disease	Fungicide and Formulation[1]	Amount of Formulation (oz/1,000 sq ft)[2]	Application Interval (days)[3]
Red Thread (*Laetisaria fuciformis*) (continued)	chlorothalonil*		
	(Daconil Ultrex) 82.5 WDG	1.8 to 3.25	7 to 10
		3.25 to 5	14
	(Daconil Weather Stik, Legend) 6 F	2 to 5.5	7 to 14
		5.5	14
	(Daconil Zn) 4.16 F	3 to 5	7 to 10
		5.3 to 8	14
	(Chlorothalonil 500ZN) 4.17 F	3 to 5	7 to 10
		7.9	14
	(Chlorothalonil 720SFT) 6 F	2.12 to 3.5	7 to 10
		5.5	14
	(Chlorothalonil, Chlorostar) 82.5 DF	1.8 to 3.2	7 to 10
	(Pegasus) 6 L	3.6 to 5.5	7 to 14
	(Pegasus) 82.5 DF	3.25 to 5	7 to 14
	(Pegasus HPX) 6 F	3.6 to 5.5	7 to 14
	chlorothalonil + acibenzolar-S-methyl (Daconil Action) 6.1 F*	2 to 3.5	7 to 10
		3.6 to 5.4	14
	chlorothalonil + fluoxastrobin (Fame C) 4.25 SC*	3 to 5.9	14 to 28
	chlorothalonil + iprodione + thiophanate-methyl + tebuconazole (Enclave) 5.3 F*	3 to 4	14 to 21
		7 to 8	28
	chlorothalonil + propiconazole (Concert) 4.3 SC*	3 to 5.5	7 to 14
		5.5 to 8.5	14 to 21
	chlorothalonil + propiconazole + fludioxonil (Instrata) 3.6 SC*	2.75 to 6	14 to 21
	chlorothalonil + thiophanate-methyl*		
	(Consyst) 67 WDG	3 to 8	7 to 10
	(Peregrine) 67 WDG	3 to 8	14
	(Spectro) 90 WDG	3.72 to 5.76	14
	(TM/C) 67 WDG	3 to 8	14 to 21
	cyazofamid + azoxystrobin (Union) 0.79 SC	2.9 to 5.75	14 to 28
	fenarimol (Rubigan) 1 AS*	8	30
	fluazinam (Secure) 4.17 SC*	0.5	14
	fluazinam + acibenzolar-S-methyl (Secure Action) 4.18 SC*	0.5	14
	fluazinam + tebuconazole (Traction) 3.24 SC*	1.3	14
	fluopyram + trifloxystrobin (Exteris Stressgard) 0.27 SC	1.5 to 4.135	14 to 28
	fluoxastrobin (Fame)		
	4 SC	0.18 to 0.36	14 to 28
	0.25 G	2.3 to 4.6 lbs	14 to 28
	fluoxastrobin + myclobutanil (Fame M) 3.9 SC	0.25 to 1	14 to 28
	fluoxastrobin + tebuconazole (Fame T) 4 SC*	0.45 to 0.9	21 to 28
	flutolanil		
	(Prostar) 70 WP, 70 WDG	1.5	21 to 28
	(Pedigree) 3.8 SC	2.2	21 to 28
	flutolanil + thiophanate-methyl (SysStar) 80 WDG	2 to 3	14 to 21
	iprodione (iprodione (26GT, Iprodione Pro, IPro, Raven) 2 F, 2 SC, 2 SE*	4	14
	iprodione + thiophanate-methyl (26/36) 3.8 F*	2 to 4	14 to 21
	iprodione + trifloxystrobin (Interface) 2.27 SC*	3 to 4	14
	isofetamid + tebuconazole (Tekken) 1.8 SC*	3	14 to 28
	mancozeb*		
	(Fore) 80 WP	4 to 8	7 to 14
	(Dithane) 75 DF	4 to 8	10
	(Pentathlon) 4 LF	7 to 14	7 to 14
	(Pentathlon) 75 DF	4 to 8	7
	(Protect, Wingman) 75 W	4 to 8	7 to 14
	mancozeb + copper hydroxide (Junction) 60 DF*	2 to 4	7 to 14
	mefentrifluconazole + pyraclostrobin (Navicon) 3.34 SC*	0.7 to 0.85	14 to 28

Turfgrass Disease Control (continued)

Disease	Fungicide and Formulation[1]	Amount of Formulation (oz/1,000 sq ft)[2]	Application Interval (days)[3]
Red Thread	metconazole (Tourney) 50 WDG	0.37	14
(*Laetisaria fuciformis*)	mineral oil (Civitas) + proprietary pigment (Civitas Harmonizer)*	(8 to 32) + (1 to 4)	7 to 21
(continued)	myclobutanil (Eagle, Myclobutanil) 20 EW	1.2	14 to 21
	penthiopyrad (Velista) 50 WG	0.3 to 0.5	14
	polyoxin D		
	(Affirm) 11.3 WDG	0.88	7 to 14
	(Endorse) 2.5 WP	4	7 to 14
	propiconazole (Banner MAXX, Kestrel, Propiconazole, Savvi, Strider) 1 ME	2	14 to 21
	pydiflumetofen + azoxystrobin + propiconazole (Posterity XT) 1.48 SE*	1.5 to 3	14 to 28
	pyraclostrobin (Insignia)		
	20 WG	0.5 to 0.9	14 to 28
	2 SC	0.4 to 0.7	14 to 28
	pyraclostrobin + boscalid (Honor) 28 WG*	0.55 to 1.1	14 to 28
	pyraclostrobin + fluxapyroxad (Lexicon Intrinsic) 4.17 SC	0.34 to 0.47	14
	pyraclostrobin + triticonazole (Pillar) 0.81 G	3 lbs	14 to 28
	tebuconazole*		
	(Torque) 3.6 F	0.6 to 1.1	refer to label
	(Mirage Stressgard) 2 SC	1 to 2	14 to 28
	(Skylark, Tebuconazole) 3.6 F	0.6	28
	thiophanate-methyl		
	(3336) 50WP or 4 F	2 to 4	14
	(3336 Plus) 2 F	2 to 4	14 to 28
	(SysTec 1998, T-Bird, TM) 85 WDG	0.67 to 1.3	14
	(3336) 2 G	1.5 to 6 lbs	14
	(SysTec 1998, T-Bird, TM) 4.5 L	1 to 2	14
	thiram (Spotrete) 4 F*	3.75 to 7.5	3 to 10
	triadimefon (Bayleton) 50 WSP, 4.15 F	0.5 to 1	15 to 30
	trifloxystrobin (Compass) 50 WDG	0.1 to 0.15	14
		0.2 to 0.25	21
	trifloxystrobin + triadimefon		
	(Armada) 50 WP	0.6 to 1.2	14 to 28
	(Tartan) 2 SC*	1 to 2	14 to 28
	triticonazole		
	(Trinity) 1.7 SC	0.5 to 1	14 to 28
	(Triton) 70 WDG	0.15 to 0.3	14 to 28
	triticonazole + chlorothalonil (Reserve) 4.79 SC*	3.2 to 4.5	refer to label
	vinclozolin (Curalan, Touche) 50 EG*	1	14 to 28
Rust	azoxystrobin		
(*Puccinia ssp.*)	(Heritage, Strobe) 50 WG	0.2 to 0.4	14 to 28
	(Heritage) 0.8 TL	1 to 2	14 to 28
	(Heritage) 0.31 G	2 to 4 lbs	14 to 28
	azoxystrobin + acibenzolar-S-methyl (Heritage Action) 51 WG*	0.2 to 0.4	14 to 28
	azoxystrobin + chlorothalonil (Renown) 5.16 SC*	2.5 to 4.5	14 to 21
	azoxystrobin + difenoconazole (Briskway) 2.7 SC*	0.5 to 1.2	14 to 28
	azoxystrobin + propiconazole (Headway)		
	1.4 ME	1.5 to 3	14 to 28
	1.06 G	2 to 4 lbs	14 to 28
	azoxystrobin + tebuconazole (Strobe T) 2.67 SC*	0.75 to 1.5	14 to 21
	benzovindiflupyr + difenoconazole (Ascernity) 0.86 SL*	1	14 to 21
	boscalid + chlorothalonil (Encartis) 6.25 SC*	4	14

Turfgrass Disease Control (continued)

Disease	Fungicide and Formulation[1]	Amount of Formulation (oz/1,000 sq ft)[2]	Application Interval (days)[3]
Rust	chlorothalonil*		
(*Puccinia ssp.*)	(Daconil Ultrex) 82.5 WDG	3.7 to 5	14
(continued)	(Daconil Weather Stik, Legend) 6 F	4.0 to 5.5	14
	(Daconil Zn) 4.16 F	6 to 8	14
	(Chlorothalonil 500ZN) 6 F	3 to 5	7 to 14
		7.9	14
	(Chlorothalonil 720SFT) 6 F	2.12 to 3.5	7 to 10
		5.5	14
	(Chlorothalonil, Chlorostar) 82.5 DF	3.2	7 to 14
	(Pegasus) 6 L	3.6 to 5.5	7 to 14
	(Pegasus) 82.5 DF	3.25 to 5	7 to 14
	(Pegasus HPX) 6 F	3.6 to 5.5	7 to 14
	chlorothalonil + acibenzolar-S-methyl (Daconil Action) 6.1 F*	4 to 5.4	14
	chlorothalonil + fluoxastrobin (Fame C) 4.25 SC*	3 to 5.9	14 to 28
	chlorothalonil + propiconazole (Concert) 4.3 SC*	3 to 5.5	7 to 14
		4.5 to 8.5	14 to 28
	chlorothalonil + propiconazole + fludioxonil (Instrata) 3.6 SC*	2.75 to 6	14 to 28
	chlorothalonil + thiophanate-methyl*		
	(Consyst) 67 WDG	3 to 8	7 to 14
	(Peregrine) 67 WDG	3 to 8	14
	(Spectro) 90 WDG	3.72 to 5.76	14
	(TM/C) 67 WDG	3 to 8	14 to 21
	cyazofamid + azoxystrobin (Union) 0.79 SC	2.9 to 5.75	14 to 28
	fluazinam (Secure) 4.17 SC*	0.5	14
	fluazinam + acibenzolar-S-methyl (Secure Action) 4.18 SC*	0.5	14
	fluazinam + tebuconazole (Traction) 3.24 SC*	1.3	14
	fluopyram + trifloxystrobin (Exteris Stressgard) 0.27 SC	1.5 to 4.135	14 to 28
	fluoxastrobin (Fame)		
	4 SC	0.18 to 0.36	14 to 28
	0.25 G	2.3 to 4.6 lbs	14 to 28
	fluoxastrobin + myclobutanil (Fame M) 3.9 SC	0.25 to 1	14 to 28
	flutriafol (Rayora) 1.04 L*	0.7 to 1.4	14 to 21
	iprodione + trifloxystrobin (Interface) 2.27 SC*	3 to 5	refer to label
	isofetamid + tebuconazole (Tekken) 1.8 SC*	3	14 to 28
	mancozeb*		
	(Fore) 80 WP	4	7 to 14
	(Dithane) 75 DF	4	10
	(Pentathlon) 4 LF	5 to 7	7 to 10
	(Pentathlon) 75 DF	4	7 to 10
	(Wingman) 75 WP	4	7 to 10
	mancozeb + copper hydroxide (Junction) 60 DF*	2 to 4	7 to 14
	mandestrobin (Pinpoint) 4SC	0.31	14
	mefentrifluconazole + pyraclostrobin (Navicon) 3.34 SC*	0.7 to 0.85	14 to 28
	metconazole (Tourney) 50 WDG	0.37	14
	myclobutanil (Eagle, Myclobutanil, Siskin) 20 EW	1.2	14 to 28
	propiconazole (Banner Maxx, Kestrel, Propiconazole, Savvi, Strider) 1 ME	1 to 2	14 to 28
	pydiflumetofen + azoxystrobin + propiconazole (Posterity XT) 1.48 SE*	1.5 to 3	14 to 28
	pyraclostrobin (Insignia)		
	20 WG	0.5 to 0.9	14 to 28
	2 SC	0.4 to 0.7	14 to 28
	pyraclostrobin + boscalid (Honor) 28WG*	0.55 to 1.1	14 to 28
	pyraclostrobin + fluxapyroxad (Lexicon Intrinsic) 4.17 SC	0.34 to 0.47	14 to 28
	pyraclostrobin + triticonazole (Pillar) 0.81 G	3 lbs	14 to 28

Turfgrass Disease Control (continued)

Disease	Fungicide and Formulation[1]	Amount of Formulation (oz/1,000 sq ft)[2]	Application Interval (days)[3]
Rust	tebuconazole*		
(*Puccinia ssp.*)	(Torque) 3.6 F	0.6 to 1.1	refer to label
(continued)	(Mirage Stressgard) 2 SC	1 to 2	14 to 28
	(Skylark, Tebuconazole) 3.6 F	0.6	28
	thiophanate-methyl		
	(3336) 50WP or 4 F	4 to 6	14
	(3336 Plus) 2 F	4 to 8	14 to 28
	(T-Bird) 4.5 L	3.5 to 5	14
	(SysTec 1998, T-Bird, TM) 85 WDG	2.35 to 3.53	14
	thiram (Spotrete) 4 F*	3.75 to 7.5	3 to 10
	triadimefon (Bayleton) 50 WSP, 4.15 F	0.5 to 1	15 to 30
	trifloxystrobin (Compass) 50 WDG	0.1 to 0.15	14
		0.2 to 0.25	21
	trifloxystrobin + triadimefon		
	(Armada) 50 WP	0.6 to 1.2	14 to 28
	(Tartan) 2 SC*	1 to 2	14 to 28
	triticonazole		
	(Trinity) 1.7 SC	0.5 to 1	14 to 28
	(Triton) 70 WDG	0.15 to 0.225	14 to 28
	triticonazole + chlorothalonil (Reserve) 4.79 SC*	3.2 to 4.5	14 to 28
Slime Mold	mancozeb (Fore) 80 WP*	4 to 8	7 to 14
(*Myxomycetes spp.*)	mancozeb + copper hydroxide (Junction) 60 DF*	2 to 4	7 to 14
Southern Blight	azoxystrobin		
(*Sclerotium rolfsii*)	(Heritage, Strobe) 50 WG	0.2 to 0.4	14 to 28
	(Heritage) 0.8 TL	1 to 2	14 to 28
	(Heritage) 0.31 G	2 to 4 lbs	14 to 28
	(Strobe) 2 L	0.38 to 0.77	14 to 28
	azoxystrobin + acibenzolar-S-methyl (Heritage Action) 51 WG*	0.2 to 0.4	14 to 28
	azoxystrobin + chlorothalonil (Renown) 5.16 SC*	2.5 to 4.5	14 to 21
	azoxystrobin + difenoconazole (Briskway) 2.7 SC*	0.5 to 1.2	14 to 28
	azoxystrobin + propiconazole (Headway)		
	1.4 ME	1.5 to 3	14 to 28
	1.06 G	2 to 4 lbs	14 to 28
	azoxystrobin + tebuconazole (Strobe T) 2.67 SC*	0.75 to 1.5	14 to 21
	benzovindiflupyr + difenoconazole (Ascernity) 0.86 SL*	1	14
	chloroneb* (Teremec) 65 SP	4	5 to 7
	chlorothalonil + fluoxastrobin (Fame C) 4.25 SC*	3 to 5.9	14 to 28
	fluoxastrobin (Fame)		
	4 SC	0.18 to 0.36	14 to 28
	0.25 G	2.3 to 4.6 lbs	14 to 28
	fluoxastrobin + myclobutanil (Fame M) 3.9 SC	0.25 to 1	14 to 28
	fluoxastrobin + tebuconazole (Fame T) 4 SC*	0.45 to 0.9	21 to 28
	flutolanil		
	(Prostar) 70 WP, 70 WDG	1.5	21 to 28
	(Pedigree) 3.8 SC	2.2	21 to 28
	flutolanil + thiophanate-methyl (SysStar) 80 WDG	2	21 to 28
	pydiflumetofen + azoxystrobin + propiconazole (Posterity XT) 1.48 SE*	1.5 to 3	14 to 28
	triadimefon (Bayleton) 50 WSP, 4.15 F	0.5 to 2	14 to 28
	trifloxystrobin + triadimefon		
	(Armada) 50 WP	0.6 to 1.2	14
	(Tartan) 2 SC*	1 to 2	14
Spring Dead Spot	azoxystrobin		
(*Ophiosphaerella korrae;*	(Heritage, Strobe) 50 WG	0.4	14 to 28
O. herpotricha; O. narmari)	(Heritage) 0.8 TL	2	14 to 28
	(Strobe) 2L	0.38 to 0.77	28

Turfgrass Disease Control (continued)

Disease	Fungicide and Formulation[1]	Amount of Formulation (oz/1,000 sq ft)[2]	Application Interval (days)[3]
Spring Dead Spot (*Ophiosphaerella korrae*; *O. herpotricha*; *O. narmari*) (continued)	azoxystrobin + acibenzolar-S-methyl (Heritage Action) 51 WG*	0.2 to 0.4	14 to 28
	azoxystrobin + propiconazole (Headway)		
	1.4 ME	3	14 to 28
	1.06 G	2 to 4 lbs	14 to 28
	azoxystrobin + tebuconazole (Strobe T) 2.67 SC*	1.5	14 to 21
	benzovindiflupyr + difenoconazole (Ascernity) 0.86 SL*	1	14 to 28
	chlorothalonil + fluoxastrobin (Fame C) 4.25 SC*	5.9	14 to 28
	chlorothalonil + iprodione + thiophanate-methyl + tebuconazole (Enclave) 5.3 F*	3 to 4	14 to 21
		7 to 8	28
	fenarimol (Rubigan) 1 AS*	4	14 to 30 (2 applications)
		6	1 application
	fluoxastrobin (Fame)		
	4 SC	0.36	14 to 28
	0.25 G	2.3 to 4.6 lbs	14 to 28
	fluoxastrobin + myclobutanil (Fame M) 3.9 SC	0.5 to 1	14 to 28
	fluoxastrobin + tebuconazole (Fame T) 4 SC*	0.45 to 0.9	21 to 28
	flutriafol (Rayora) 1.04 L*	0.7 to 1.4	21 to 28
	isofetamid (Kabuto) 3.33 SC	0.5 to 1.6	14 to 28
	isofetamid + tebuconazole (Tekken) 1.8 SC*	3	14 to 28
	mefentrifluconazole (Maxtima) 3.34 SC*	0.6 to 0.8	28
	mefentrifluconazole + pyraclostrobin (Navicon) 3.34 SC*	0.85	28
	myclobutanil (Eagle, Myclobutanil, Siskin) 20 EW	2.4	28 (fall)
	penthiopyrad (Velista) 50 WG	0.7	28
	propiconazole (Banner MAXX, Kestrel, Propiconazole, Savvi, Strider) 1 ME	4	30
	prothioconazole (Densicor) 4 SC*	0.196	14 to 28
	pydiflumetofen (Posterity) 1.67 SC*	0.16 to 0.32	28
	pydiflumetofen + azoxystrobin + propiconazole (Posterity Forte) 2.5 SE*	0.63 to 0.84	14 to 28
	pydiflumetofen + azoxystrobin + propiconazole (Posterity XT) 1.48 SE*	3	28
	tebuconazole*		
	(Torque) 3.6 F	0.6 to 1.1	21
	(Mirage Stressgard) 2 SC	2	28
	(Skylark, Tebuconazole) 3.6 F	0.6	fall and spring
	thiophanate-methyl		
	(3336) 50WP or 4 F	4 to 6	14
	(3336) 2 G	6 to 9 lbs	14
Stripe Smut (*Ustilago striiformis*)	chlorothalonil + iprodione + thiophanate-methyl + tebuconazole (Enclave) 5.3 F*	3 to 4	14 to 21
		7 to 8	28
	chlorothalonil + propiconazole (Concert) 4.3 SC*	4.5 to 8.5	fall or spring
	fluazinam + tebuconazole (Traction) 3.24 SC*	1.3	one application
	isofetamid + tebuconazole (Tekken) 1.8 SC*	3	14 to 28
	myclobutanil (Eagle, Myclobutanil, Siskin) 20 EW	1.2	14
	propiconazole (Banner MAXX, Kestrel, Propiconazole, Savvi, Strider) 1 ME	1 to 2	fall or spring
	tebuconazole*		
	(Torque) 3.6 F	0.6 to 1.1	spring
	(Skylark, Tebuconazole) 3.6 F	0.6	spring
	thiophanate-methyl		
	(3336) 50WP or 4 F	4 to 6	14
	(3336 Plus) 2 F	4 to 8	14 to 28
	(3336) 2 G	6 to 9 lbs	14
	(T-Bird) 4.5 L	5 to 10	14 to 21
	(SysTec 1998, T-Bird, TM) 85 WDG	3 to 3.53	14 to 21
	(SysTec 1998, T-Bird,TM) 4.5 L	5	14 to 21
	triadimefon (Bayleton) 50 WSP	1	refer to label

Turfgrass Disease Control (continued)

Disease	Fungicide and Formulation[1]	Amount of Formulation (oz/1,000 sq ft)[2]	Application Interval (days)[3]
Stripe Smut	trifloxystrobin + triadimefon		
(*Ustilago striiformis*)	(Armada) 50 WP	0.6	refer to label
(continued)	(Tartan) 2 SC*	1	refer to label
Summer Patch	azoxystrobin		
(*Magnaporthe poae*)	(Heritage, Strobe) 50 WG	0.2 to 0.4	14 to 28
	(Heritage) 0.8 TL	1 to 2	14 to 28
	(Heritage) 0.31 G	2 to 4 lbs	14 to 28
	(Strobe) 2 L	0.38 to 0.77	14 to 28
	azoxystrobin + acibenzolar-S-methyl (Heritage Action) 51 WG*	0.2 to 0.4	14 to 28
	azoxystrobin + difenoconazole (Briskway) 2.7 SC*	0.5 to 1.2	14 to 28
	azoxystrobin + propiconazole (Headway)		
	1.4 ME	1.5 to 3	14 to 28
	1.06 G	2 to 4 lbs	14 to 28
	benzovindiflupyr + difenoconazole (Ascernity) 0.86 SL*	1	14
	chlorothalonil + fluoxastrobin (Fame C) 4.25 SC*	3 to 5.9	14 to 28
	chlorothalonil + iprodione + thiophanate-methyl + tebuconazole	3 to 4	14 to 21
	(Enclave) 5.3 F*	7 to 8	28
	chlorothalonil + propiconazole + fludioxonil (Instrata) 3.6 SC*	6 to 11	14 to 28
	cyazofamid + azoxystrobin (Union) 0.79 SC	2.9 to 5.75	14 to 28
	fenarimol (Rubigan) 1 AS*	2 to 4	30 (2 applications)
		2	30 (greens)
		4 to 8	single application
	fluoxastrobin (Fame)		
	4 SC	0.18 to 0.36	14 to 28
	0.25 G	2.3 to 4.6 lbs	14 to 28
	fluoxastrobin + myclobutanil (Fame M) 3.9 SC	0.25 to 1	14 to 28
	fluoxastrobin + tebuconazole (Fame T) 4 SC*	0.45 to 0.9	21 to 28
	fludioxonil (Medallion) 50 WP	0.5	14
	flutriafol (Rayora) 1.04 L*	0.7 to 1.4	21 to 28
	fluxapyroxad (Xzemplar) 2.47 SC	0.26	14 to 28
	isofetamid + tebuconazole (Tekken) 1.8 SC*	3	14 to 28
	mefentrifluconazole (Maxtima) 3.34 SC*	0.8	21 to 28
	mefentrifluconazole + pyraclostrobin (Navicon) 3.34 SC*	0.7 to 0.85	14 to 28
	metconazole (Tourney) 50 WDG	0.37	14
	myclobutanil (Eagle, Myclobutanil, Siskin) 20 EW	1.2 to 2.4	14 to 28
	penthiopyrad (Velista) 50 WG	0.3 to 0.5	14 to 28
	propiconazole (Banner MAXX, Kestrel, Propiconazole, Savvi, Strider) 1	2	14
	ME	4	28
	prothioconazole (Densicor) 4 SC*	0.196	14 to 28
	pydiflumetofen + azoxystrobin + propiconazole (Posterity XT) 1.48 SE*	1.5 to 3	14 to 28
	pyraclostrobin (Insignia)		
	20 WG	0.5 to 0.9	14 to 28
	2 SC	0.4 to 0.7	14 to 28
	pyraclostrobin + boscalid (Honor) 28 WG*	1.1	14 to 28
	pyraclostrobin + fluxapyroxad (Lexicon Intrinsic) 4.17 SC	0.34 to 0.47	14 to 28
	pyraclostrobin + triticonazole (Pillar) 0.81 G	3 lbs	28
	tebuconazole*		
	(Torque) 3.6 F	0.6 to 1.1	21
	(Mirage Stressgard) 2 SC	1 to 2	14 to 28
	(Skylark, Tebuconazole) 3.6 F	0.6	28

Turfgrass Disease Control (continued)

Disease	Fungicide and Formulation[1]	Amount of Formulation (oz/1,000 sq ft)[2]	Application Interval (days)[3]
Summer Patch (*Magnaporthe poae*) (continued)	thiophanate-methyl		
	(3336) 50WP or 4 F	4 to 6	14 to 21
	(3336 Plus) 2 F	4 to 8	14 to 28
	(3336) 2 G	6 to 9 lbs	14 to 21
	(SysTec 1998, T-Bird, TM) 85 WDG	3.53	14
	(SysTec 1998, T-Bird, TM) 4.5 L	5	14
	triadimefon (Bayleton) 50 WSP, 4.15 F	1 to 2	30
	trifloxystrobin (Compass) 50 WDG	0.2 to 0.25	21 to 28
	trifloxystrobin + triadimefon		
	(Tartan) 2 SC *	2	21 to 28
	(Armada) 50 WP	1.2	21 to 28
	triticonazole		
	(Trinity) 1.7 SC	1 to 2	14 to 28
	(Triton) 70 WDG	0.3 to 0.6	14 to 28
	triticonazole + chlorothalonil (Reserve) 4.79 SC*	3.2 to 5.4	14 to 28
Take-All Patch (*Gaeumannomyces avenae*)	azoxystrobin		
	(Heritage, Strobe) 50 WG	0.4	28
	(Heritage) 0.8 TL	2	28
	(Heritage) 0.31 G	2 to 4 lbs	28
	(Strobe) 2 L	0.38 to 0.77	28
	azoxystrobin + acibenzolar-S-methyl (Heritage Action) 51 WG*	0.2 to 0.4	28
	azoxystrobin + difenoconazole (Briskway) 2.7 SC*	0.5 to 1.2	28
	azoxystrobin + propiconazole (Headway)		
	1.4 ME	3	14 to 28
	1.06 G	2 to 4 lbs	14 to 28
	benzovindiflupyr + difenoconazole (Ascernity) 0.86 SL*	1	14
	chlorothalonil + fluoxastrobin (Fame C) 4.25 SC*	5.9	28
	cyazofamid + azoxystrobin (Union) 0.79 SC	5.75	28
	fenarimol (Rubigan) 1 AS*	4	30 (greens)
		4 to 8	30 (1 or 2 applications)
	fluoxastrobin (Fame)		
	4 SC	0.36	28
	0.25 G	2.3 to 4.6 lbs	28
	fluoxastrobin + myclobutanil (Fame M) 3.9 SC	0.5 to 1	28
	fluoxastrobin + tebuconazole (Fame T) 4 SC*	0.45 to 0.9	28
	isofetamid + tebuconazole (Tekken) 1.8 SC*	3	14 to 28
	mandestrobin (Pinpoint) 4SC	0.31	14
	mefentrifluconazole (Maxtima) 3.34 SC*	0.8	28
	mefentrifluconazole + pyraclostrobin (Navicon) 3.34 SC*	0.7 to 0.85	14 to 28
	myclobutanil (Eagle, Myclobutanil, Siskin) 20 EW	2.4	28 (spring/fall)
	propiconazole (Banner MAXX, Kestrel, Propiconazole, Savvi, Strider) 1 ME	2 to 4	spring and fall
	pydiflumetofen + azoxystrobin + propiconazole (Posterity XT) 1.48 SE*	3	21
	pyraclostrobin (Insignia)		
	20 WG	0.9	28
	2 SC	0.7	28
	pyraclostrobin + boscalid (Honor) 28 WG*	1.1	28
	pyraclostrobin + fluxapyroxad (Lexicon Intrinsic) 4.17 SC	0.47	28
	pyraclostrobin + triticonazole (Pillar) 0.81 G	3 lbs	28
	tebuconazole*		
	(Torque) 3.6 F	0.6 to 1.1	fall and spring
	(Mirage Stressgard) 2 SC	1 to 2	14 to 28
	(Skylark, Tebuconazole) 3.6 F	0.6	fall and spring

Turfgrass Disease Control (continued)

Disease	Fungicide and Formulation[1]	Amount of Formulation (oz/1,000 sq ft)[2]	Application Interval (days)[3]
Take-All Patch (*Gaeumannomyces avenae*) (continued)	thiophanate-methyl		
	(3336) 50 WP or 4 F	4 to 6	14
	(3336 Plus) 2 F	4 to 8	14 to 28
	(3336) 2 G	6 to 9 lbs	14
	triadimefon (Bayleton) 50 WSP, 4.15 F	1 to 2	21 to 28
	trifloxystrobin + triadimefon (Armada) 50 WP	1.2	28
	triticonazole		
	(Trinity) 1.7 SC	1 to 2	14 to 28
	(Triton) 70 WDG	0.15 to 0.3	14 to 28
	triticonazole + chlorothalonil (Reserve) 4.79 SC*	3.2 to 5.4	14 to 28
Take-all Root Rot/ Bermudagrass Decline (*Gaeumannomyces graminis*)	azoxystrobin + acibenzolar-S-methyl (Heritage Action) 51 WG*	0.4	28
	azoxystrobin + difenoconazole (Briskway) 2.7 SC*	0.5 to 1.2	14 to 28
	isofetamid + tebuconazole (Tekken) 1.8 SC*	3	14 to 28
	fluxapyroxad + pyraclostrobin (Lexicon Intrinsic) 4.17 SC	0.34 to 0.47	refer to label
	mefentrifluconazole (Maxtima) 3.34 SC*	0.8	28
	mefentrifluconazole + pyraclostrobin (Navicon) 3.34 SC*	0.7 to 0.85	14 to 28
	prothioconazole (Densicor) 4 SC*	0.196	14 to 28
	pydiflumetofen + azoxystrobin + propiconazole (Posterity XT) 1.48 SE*	1.5 to 3	14 to 28
	pyraclostrobin (Insignia)		
	20 WG	0.9	refer to label
	2 SC	0.7	refer to label
	pyraclostrobin + boscalid (Honor) 28 WG*	1.1	refer to label
	pyraclostrobin + triticonazole (Pillar) 0.81G	3 lbs	28
	tebuconazole*		
	(Torque) 3.6 F	0.6 to 1.1	14 to 28
	(Mirage Stressgard) 2SC	2	28
	thiophanate-methyl		
	(3336) 50 WP or 4 F	4 to 6	14
	(3336 Plus) 2 F	4 to 8	14 to 28
	(3336) 2 G	6 to 9 lbs	14
	triadimefon (Bayleton) 50 WSP, 4.15 F	1 to 2	21 to 28
Yellow Patch (*Rhizoctonia cerealis*)	azoxystrobin		
	(Heritage, Strobe) 50 WG	0.4	28
	(Heritage) 0.8 TL	2	28
	(Heritage) 0.31 G	2 to 4 lbs	14 to 28
	(Strobe) 2 L	0.38 to 0.77	28
	azoxystrobin + acibenzolar-S-methyl (Heritage Action) 51 WG*	0.2 to 0.4	14 to 28
	azoxystrobin + chlorothalonil (Renown) 5.16 SC*	2.5 to 4.5	14 to 28
	azoxystrobin + difenoconazole (Briskway) 2.7 SC*	0.5 to 1.2	14 to 28
	azoxystrobin + propiconazole (Headway)		
	1.4 ME	3	28
	1.06 G	2 to 4 lbs	14 to 28
	azoxystrobin + tebuconazole (Strobe T) 2.67 SC*	1.5	14 to 21
	benzovindiflupyr + difenoconazole (Ascernity) 0.86 SL*	1	14 to 21
	chlorothalonil + fluoxastrobin (Fame C) 4.25 SC*	3 to 5.9	14 to 28
	chlorothalonil + propiconazole + fludioxonil (Instrata) 3.6 SC*	8 to 11	late fall
	chlorothalonil + thiophanate-methyl (Spectro) 90 WDG*	3 to 5.76	14 to 21
	cyazofamid + azoxystrobin (Union) 0.79 SC	5,75	28
	fludioxonil (Medallion) 50 WP	0.5	1 application
	fluopyram + trifloxystrobin (Exteris Stressgard) 0.27 SC	2.135 to 6	21 to 28
	fluoxastrobin (Fame)		
	4 SC	0.36	28
	0.25 G	2.3 to 4.6 lbs	14 to 28
	fluoxastrobin + myclobutanil (Fame M) 3.9 SC	0.25 to 1	28
	fluoxastrobin + tebuconazole (Fame T) 4 SC*	0.45 to 0.9	21 to 28

Turfgrass Disease Control (continued)

Disease	Fungicide and Formulation[1]	Amount of Formulation (oz/1,000 sq ft)[2]	Application Interval (days)[3]
Yellow Patch	flutolanil		
(*Rhizoctonia cerealis*)	(Prostar) 70 WP, 70 WDG	1.5	21 to 28
(continued)	(Pedigree) 3.8 SC	2.2	21 to 28
	flutolanil + thiophanate-methyl (SysStar) 80 WDG	2	21 to 28
	isofetamid + tebuconazole (Tekken) 1.8 SC*	3	14 to 28
	metconazole (Tourney) 50 WDG	0.37 to 0.44	late fall
	polyoxin D		
	(Affirm) 11.3 WDG	0.88	7 to 14
	(Endorse) 2.5 WP	4	7 to 14
	propiconazole (Banner MAXX, Kestrel, Propiconazole, Savvi, Strider) 1 ME	3 to 4	late fall
	pydiflumetofen + azoxystrobin + propiconazole (Posterity XT) 1.48 SE*	3	28
	tebuconazole (Mirage Stressgard) 2 SC*	1 to 2	21 to 28
	thiophanate-methyl		
	(3336) 50WP or 4 F	4 to 6	14
	(3336 Plus) 2 F	4 to 8	14 to 28
	(3336) 2 G	6 to 9 lbs	14
	triticonazole		
	(Triton FLO) 3 F	0.55 to 1.1	21 to 28
	(Trinity) 1.75 SC	1 to 2	21 to 28
	triticonazole + chlorothalonil (Reserve) 4.79 SC*	3.2 to 5.4	21 to 28
Yellow Tuft	fosetyl Al		
(*Sclerophthora macrospora*)	(Signature, Fosetyl-Al) 80 WDG	4 to 8	14 to 21
	(Signature Xtra Stressgard) 60 WDG*	2 to 6	14 to 21
	(Autograph) 70 DF*	4.6 to 9.2	14 to 21
	(Viceroy) 70 DF	4.6 to 9.1	14 to 21
	mefenoxam		
	(Subdue WSP) 43 WSP	0.28 to 0.56	10 to 21
	(Subdue Maxx, Quell 2 ME	0.5 to 1	10 to 21
	(Subdue GR) 1 G	12.5 to 25	10 to 14
	(Mefenoxam, Fenox) 2 AQ, 2 EC	0.2 to 1	10 to 21
	mefentrifluconazole + pyraclostrobin (Navicon) 3.34 SC*	0.7 to 0.85	14 to 28
	metalaxyl (Vireo) 2 MEC	1 to 2	10 to 21
	pyraclostrobin (Insignia)		
	20 WG	0.5 to 0.9	14 to 28
	2 SC	0.4 to 0.7	14 to 28
	pyraclostrobin + boscalid (Honor) 28 WG*	0.55 to 1.1	14 to 28
	pyraclostrobin + fluxapyroxad (Lexicon Intrinsic) 4.17 SC	0.34 to 0.47	14 to 28
	pyraclostrobin + triticonazole (Pillar) 0.81 G	3 lbs	14 to 28
Zoysia Patch	See Large Patch		

[1]Other trade names with the same active ingredients are labeled for use on turfgrasses and can be used according to label directions.

[2]Apply fungicides in 2 to 5 gallons of water per 1,000 square feet according to label directions. Use lower rates for preventive and higher rates for curative applications.

[3]Use shorter intervals when conditions are very favorable for disease.

*Products marked with an asterisk are not labeled for home lawn use.

Nematicides for Turf

J. P. Kerns and E. L. Butler, Entomology and Plant Pathology Extension

Nematicides for Turf

Nematicide and Formulation	Amount of Formulation Per 1,000 sq ft	Precautions and Remarks
abamectin (Avid) 0.15 EC	1.31	For use on golf course greens only. Only abamectin formulated as Avid can be used for nematode control in turf according to a 24(c) label. Apply Avid 0.15 EC as an early curative treatment (after appropriate nematode extraction, identification, and counts). Apply early in the morning while grass is wet with dew or irrigate the area prior to application with 0.1 inch of water. Immediately after application irrigate with 0.1 inch of water to move treatments through the thatch. Do not over irrigate. Apply 3 to 4 consecutive Avid 0.15 EC applications on a 14- to 21-day interval. Avid is labeled only for sting (*Belonolaimus longicaudatus*) and ring (*Macroposthonia* sp.) nematodes.
abamectin (Divanem) 0.7 SC	3.125 to 12.2	For control of turf-parasitic nematodes on golf course greens, tees, fairways and professional and collegiate sports fields. It is not labeled for use on golf course roughs, residential turf, or commercial turf. Divanem should be applied as a seasonal program. Apply Divanem as an early curative treatment after appropriate nematode extraction, identification, and counts. Multiple applications may be required before improvements in turf quality are observed. Maximum annual rate must not exceed 0.27 lb abamectin/calendar year or 50 fl oz Divanem/A/calendar year. Do not apply to turf under heat or moisture stress. Apply in the early morning while grass is wet with dew or irrigate the area prior to application with 0.1 inches of water. Spray onto wet turf. Irrigate with 0.1 to 0.5 inches of water beginning within 1 hour of application to move Divanem through the thatch. For best results, irrigate before the spray droplets have dried on the turf. Apply in 2 gallons of water per 1,000 square feet of turf. Application rate is 3.125 to 6.25 fl oz/A every 14 to 21 days or 6.25 to 12.2 fl oz/A every 21 to 28 days.
Bacillus firmus (Nortica)	10 to 30	Nortica is a biological agent for the protection of plant roots against plant parasitic nematodes on turf, lawns, sod farms, and golf courses. Do not mix with other chemicals or fertilizers during application without first contacting a local Bayer representative. Do not apply for a month after a formaldehyde application. Do not apply within 2 weeks of a fumigant application. Do not combine in the spray tank with pesticides, surfactants or fertilizers if there has been no previous experience or use of the combination to show it is physically compatible, effective, and non-injurious under your use conditions. Refer to product label for further information about mixing compatibilities. Nortica is suitable for application by spraying, drenching, or by drip irrigation. Optimal results are obtained by pre-plant applications (from 2 to 7 days prior to planting) and immediately irrigating after application to a minimum of 3 to 4 inches. If product is applied prior to planting, maintain moist soil with daily irrigation until planting. Refer to label for further information about application techniques. Make applications every 3 months as necessary and irrigate to a depth of 4 inches.
fluensulfone (Nimitz Pro G)	1.38 to 2.75 lbs	Intended for use by a commercial applicator. For nematode control in bermudagrass, St. Augustinegrass, zoysiagrass, centipedegrass, seashore paspalum, tall fescue, and creeping bentgrass on golf courses, sports fields, commercial turfgrass areas, sod farms, and residential turfgrass lawns. Nimitz must be immediately watered in after application with a minimum of ¼ inch water. Do not make applications when soil temperature is below 55°F. Do not exceed 240 pounds of product per acre per calendar year. Do not allow bystanders to enter the treated area until the granules have been watered-in.
fluopyram (Indemnify)	0.195 to 0.39	Intended for use by commercial applicators. For use on golf courses, sod farms, sports fields, residential, institutional, municipal, commercial, and other turfgrass areas. Do not apply more than the maximum annual rate for each specific use from any combination of products containing fluopyram. Do not apply via aerial application. For ground application equipment, apply 2 to 5 gallons of solution per 1000 sq. ft. When using against nematodes, for optimum control irrigate within 24 hours of application to depth of the root zone to be protected. Do not apply more than 17.1 fl oz of Indemnify per acre per year. For residential turf, do not apply more than 15.5 fl oz of Indemnify per acre per year.
furfural (Multiguard Protect)	0.126 to 0.184	For terrestrial (outdoor) non-food use on established turf on golf course tees and greens, practice greens, spot treatment of fairways, roughs, and turf/sod farms. Areas to be treated must be at least 70% of field capacity before application. Apply up to 6 applications using only ground boom sprayers set to release spray at no more than 2 feet above the ground. Use the high rate at the start of the season and under high infestation and/or until acceptable control is achieved every 14 to 28 days. Then use the lower rate as a maintenance application at 14- to 28-day intervals.

Growth Regulators for Turfgrasses

F. H. Yelverton, R. Cooper, and T. W. Gannon, Crop and Soil Sciences Department

Growth Regulators for Turfgrasses

Turfgrass	Brand	Amount of Formulation Per Acre	Pounds Active Ingredient Per Acre	Precautions and Remarks
Cool Season Grasses—Well-Maintained Turf: Seedhead and Foliar Suppression	mefluidide (Embark) 0.2	5 pt/15 to 150 gal water	0.125	See Embark 2-S for low-maintenance cool-season turf. Follow label directions and precautions.
	trinexapac-ethyl (Governor) 0.17 G	30 to 258 lb	0.05 to 0.44	Apply 30 to 41 pounds per acre to greens, 53 to 152 pounds per acre to fairways less than 0.5 inch cut, and 152 to 258 pounds per acre to residential and commercial turf. Do not exceed 2.5 pounds active ingredient per acre per year. These rates should provide 50% turf growth suppression for 4 weeks with minimal yellowing.
	(Primo Maxx) 1 MEC or (T-Nex) 1 AQ (Primo WSB) 25 WP	6 to 44 fl oz 2.75 to 21.8 oz	0.085 to 0.34 0.085 to 0.34	Application rate varies with turfgrass species and height of cut. Apply to actively growing, nonstressed turf. More growth suppression occurs at lower mowing heights. See label for specific rate and other directions and precautions. Repeat applications can be made, but do not exceed a total of 21.4 pints per acre per year of Primo Maxx or a total of 174 ounces per acre per year of Primo WSB. Do not exceed a total of 19 pints per acre per year of T-Nex. Refer to the respective Primo label for guidelines regarding mowing prior to and following application. Mix with 0.5 to 4 gallons of water per 1,000 sq ft (20 to 174 gallons per acre). Primo can be applied to putting greens. See label for instructions.
Cool Season Grasses—Well-Maintained Turf: Foliar Suppression	ethephon (Ethephon or Proxy) 2 SL	1.7 gal	3.4	May be applied to Kentucky bluegrass, perennial ryegrass, bentgrass, and tall and fine fescues. Apply in 22 to 174 gallons of water per acre. Do not use a surfactant. Plant growth regulator effect will not be seen until 7 to 10 days after application. May be reapplied to Kentucky bluegrass and perennial ryegrass at 7-week intervals. Repeat applications to bentgrass and tall and fine fescue may be made at 4-week intervals.
	flurprimidol (Cutless 50 W) 50 WP	0.75 to 3 lb/50 to 200 gal water	0.37 to 1.5	Rates depend upon grass species and cultivar. Apply to bentgrass, Kentucky bluegrass, and perennial ryegrass in late spring-early summer and/or late summer-early fall. Time the second application to occur at least 3 months before expected winter dormancy. Do not apply to putting greens. Do not exceed 1.5 pounds per acre per application on coarse-textured soils. Treated areas should receive 0.5 inch of irrigation within 24 hours after application. Resume mowing 3 to 5 days after application.
	flurprimidol + trinexapac-ethyl (Legacy) 1.51 SL	5 to 22 fl oz	0.059 to 0.26	Tolerant species include bentgrass greens and fairways, Kentucky bluegrass, and perennial ryegrass. Do not use on turf grown for sale or other commercial use as sod or seed production. Do not seed 3 weeks before or 3 weeks after application. Wait 6 to 8 weeks after sprigging or laying sod before applying. Use only 5 to 8 fluid ounces per acre on bentgrass greens. Repeat applications at 2- to 6-week intervals until 4 weeks before the onset of inactive growth.

Growth Regulators for Turfgrasses (continued)

Turfgrass	Brand	Amount of Formulation Per Acre	Pounds Active Ingredient Per Acre	Precautions and Remarks
Cool Season Grasses—Well-Maintained Turf: Foliar Suppression (continued)	paclobutrazol (TGR Turf Enhancer 2 SC or Trimmit 2 SC) 2 SC	1 to 2 pt /43 to 200 gal water	0.25 to 0.5	Apply in spring after greenup and after turf has been mowed once or twice. Apply at least 1 month before onset of high temperatures. In late summer-early fall, apply at least 1 month before anticipated first killing frost. Apply with 0.5 to 0.9 pound nitrogen per 1000 sq ft of a nonburning fertilizer. Apply 0.25 inch of water within 24 hours after application to remove product from foliage and onto soil surface. See label for special rates and directions for applications to bentgrass, putting greens, and overseeded bermudagrass. Repeat applications within the same growing season may be made but refer to label for instructions. Do not apply more than three times annually. Do not use on areas containing greater than 70% *poa annua*. Do not seed within 6 weeks prior to or 2 weeks after applications.
	prohexadione calcium (Anuew) 27.5 WG	1.8 to 29.1 oz	0.031 to 0.5	Apply to golf course fairways, tees, greens and roughs and also athletic fields, residential and commercial lawns, sod farms, parks, cemeteries and roadsides. Apply 1.8 to 7.25 ounces per acre on bentgrass greens and tees at 1 to 2 wk intervals and 7.25 to 14.5 ounces per acre on bentgrass fairways and roughs at 2 to 4 wk intervals. Apply 14.5 to 21.8 ounces per acre on perennial bluegrass and 21.8 to 29.1 ounces per acre on perennial ryegrass at 2 to 4 wk intervals. For cool season grass sod production, apply 7.25 to 29.1 ounces per acre at 2 to 4 wk intervals. Use a spray volume of 1 to 2 gallons of water per 1000 sq ft. A nonionic surfactant may improve product performance. Do not irrigate for 4 hours after application or mow until 1 day after application.
	trinexapac-ethyl (Primo Maxx) 1 MEC or (T-Nex) 1 AQ (Primo WSB) 25 WP	6 to 22 fl oz 2.75 to 10.9 oz	0.085 to 0.17 0.085 to 0.17	Application rates are for mowing heights of less than or equal to 0.5 inch Apply to actively growing, non-stressed turf. Rate varies with turfgrass species. See label for specific rate and other directions and precautions. Repeat applications can be made but do not exceed a total of 21.4 pints per acre per year of Primo Maxx or a total of 174 ounces per acre per year of Primo WSB. Do not exceed a total of 19 pints per acre per year of T-Nex. Refer to the respective Primo label for guidelines regarding mowing prior to and following applications. Mix with 0.5 to 4 gallons of water per 1,000 sq ft (20 to 174 gallons per acre). Primo can be applied to putting greens. See label for instructions.
Cool Season Grasses—Low-Maintenance Turf: Seedhead and Foliar Suppression	chlorsulfuron (Telar DF) 75 DF + mefluidide (Embark 2-S) 2S	0.25 oz + 0.5 pt	0.012 + 0.125	For growth and seedhead suppression in fescue/bluegrass stands. Apply up until seedhead emergence. Do not apply Telar DF to turf less than 1 year old. Grass seed may be planted in treated areas 6 months after treatment, but cultivation is recommended. For broadcast applications, do not exceed 0.5 ounces Telar DF per acre within a 12-month period. Telar DF alone can also be used for weed control in bahiagrass, bermudagrass, fescue, and bluegrass.

Growth Regulators for Turfgrasses (continued)

Turfgrass	Brand	Amount of Formulation Per Acre	Pounds Active Ingredient Per Acre	Precautions and Remarks
Cool Season Grasses—Low-Maintenance Turf: Seedhead and Foliar Suppression (continued)	glyphosate (Touchdown Pro) 3 LC	4 to 8 fl oz/10 to 40 gal water	0.09375 to 0.1875	Touchdown Pro may be used on turf described in "GENERAL USE AREAS" section of the label. 4 to 5 ounces will suppress annual grasses, such as ryegrass, wild barley, and wild oats, growing in turf areas. 6 ounces will suppress Kentucky bluegrass and serve as a mowing substitute. 8 ounces will suppress fine fescue and tall fescue and serve as a mowing substitute. A nonionic surfactant containing at least 75% active ingredient at 0.25% v/v (1 quart per 100 gallons) or ammonium sulfate at 0.5% by weight (4.25 to 17 pounds per 100 gallons) may be added.
	imazethapyr + imazapyr (Event) 1.46 lb/gal	8 to 10 fl oz	0.09 to 0.11	Apply to tall fescue, perennial ryegrass, and bluegrass only. Apply after the turf is at 100% greenup and has at least 2 inches of vertical growth. The addition of a nonionic surfactant containing at least 80% active ingredient at 0.25% v/v of the spray (2 pints per 100 gallons of spray mixture) is required. Do not use on newly established stands less than 1 year old or on highly managed turf. Do not reseed before 3 months after application. See label for herbicide tank mix options. Follow label directions and precautions.
	maleic hydrazide (Retard) 2.25 lb/gal (Royal Slo-Gro) 1.5 lb/gal (Liquid Growth Retardant) 0.6 lb/gal	1.3 gal/50 gal water 2 gal/30 to 50 gal water 5 gal/45 gal water	3	Treat in the spring when the grass is actively growing but before seedhead appears. Applications made after seedhead appears will suppress subsequent seedheads. Do not apply to turf less than 3 years old, and do not reseed within 3 days after application. Treated turf may appear less dense and temporarily discolored. Optimum results may not be obtained if rainfall or overhead irrigation occurs within 12 hours following application. Remove excess grass clippings and fallen leaves before application. Do not add a surfactant. Follow label directions and precautions.
	mefluidide (Embark 2-S) 2 S	1.5 to 2 pt/15 to 150 gal water	0.38 to 0.5	Apply after uniform spring greenup until approximately 2 weeks before seedheads appear. Do not apply to turf within 4 growing months after seeding, and do not reseed within 3 days after application. Treated turf may appear less dense and temporarily discolored. Optimum results may not be obtained if rainfall or overhead irrigation occurs within 8 hours following application. Remove excess clippings and fallen leaves before application. Adding 1 to 2 quarts of nonionic surfactant per 100 gallons of spray solution may enhance suppression; however, discoloration may also be increased. Follow label directions and precautions.
	metsulfuron methyl (Escort XP) 60 DF	0.25 to 0.5 oz	0.009 to 0.018	Apply to well-established tall fescue and perennial bluegrass turf. Can tank mix with 0.125 to 0.25 pints per acre of Embark to improve pgr performance. Treat after 2 to 3 inches of new growth but before seed stalk formation. Temporary discoloration may occur. Do not use on stressed turf.
Warm Season Grasses—Well-Maintained Turf: Seedhead and Foliar Suppression	trinexapac-ethyl (Governor) 0.17 G	12 to 258 lb	0.02 to 0.44	Apply 12 to 41 pounds per acre to greens, 30 to 77 pounds per acre to fairways less than 0.5 inch cut, and 41 to 258 pounds per acre to residential and commercial turf. Do not exceed 2.5 pounds active ingredient per acre per year. These rates should provide 50% turf growth suppression for 4 weeks with minimal yellowing.

Growth Regulators for Turfgrasses (continued)

Turfgrass	Brand	Amount of Formulation Per Acre	Pounds Active Ingredient Per Acre	Precautions and Remarks
Warm Season Grasses— Well- Maintained Turf: Seedhead and Foliar Suppression (continued)	(Primo Maxx) 1 MEC or (T-Nex) 1 AQ (Primo WSB) 25 WP	2.7 to 88 fl oz 1.35 to 43.6 oz	0.085 to 0.68 0.085 to 0.68	Application rate varies with turfgrass species and height of cut. Apply to actively growing, nonstressed turf. More growth suppression occurs at lower mowing heights. See label for specific rate and other directions and precautions. Repeat applications can be made but do not exceed a total of 21.4 pints per acre per year of Primo Maxx or 174 ounces per acre per year of Primo WSB. Do not exceed a total of 19 pints per acre per year of T-Nex. Refer to the respective Primo label for guidelines regarding mowing prior to and following application. Mix with 0.5 to 4 gallons of water per 1,000 sq ft (20 to 174 gallons per acre). Primo can be applied to putting greens. See label for directions.
	mefluidide (Embark) 0.2	10 pt/15 to 150 gal water	0.25	For St. Augustinegrass. See Embark 2-S for low-maintenance warm season turf. Follow label directions and precautions.
Warm Season Grasses— Well- Maintained Turf: Foliar Suppression	flurprimidol (Cutless 50 W) 50 WP	0.75 to 3 lb/50 to 200 gal water	0.37 to 1.5	Rates depend upon grass species and cultivar. Apply to Tifway, Tifgreen, common bermudagrass, or zoysiagrass. Treated areas should receive 0.5 inch of irrigation within 24 hours of application. Resume mowing. Overseed 2 to 3 weeks after fall application with a desired perennial ryegrass.
	flurprimidol + trinexapac-ethyl (Legacy) 1.51 SL	8 to 15 fl oz	0.094 to 0.177	Tolerant species include Tifway and Tifsport bermudagrass, zoysiagrass, and seashore paspalum. Do not use on turf grown for sale or other commercial use as sod or seed production. Do not seed 3 weeks before or 3 weeks after application. Wait 6 to 8 weeks after sprigging or laying sod before applying. Repeat applications at 2- to 6-week intervals until 4 weeks before winter dormancy.
	paclobutrazol (TGR Turf Enhancer 2 SC or Trimmit 2 SC) 2 SC	2 to 3 pt/43 to 200 gal water	0.5 to 0.75	Use any time when established hybrid bermudagrass and St. Augustinegrass are green, are actively growing, and have recovered from dormancy (filled in fully following winter). Apply with 0.5 to 0.9 pound nitrogen per 1,000 sq ft of a nonburning fertilizer. Apply 0.25 inch of water within 24 hours after application to remove product from foliage and onto soil surface. A repeat application within the same growing season may be made, but not sooner than 8 weeks following initial application. Do not apply more than 3 times annually. Do not use on areas containing greater than 70% *poa annua*. Refer to label to determine bermudagrass and St. Augustine cultivar response relating to sensitivity, growth, and color response. Do not seed within 6 weeks prior to or 2 weeks after application.
	prohexadione calcium (Anuew) 27.5 WG	7.25 to 43.6 oz	0.125 to 0.75	Apply to golf course fairways, tees, greens and roughs and also athletic fields, residential and commercial lawns, sod farms, parks, cemeteries and roadsides. Apply 7.25 to 14.5 ounces per acre on hybrid bermudagrass greens and tees at 1 to 2 wk intervals and 29.1 to 43.6 ounces per acre on hybrid bermudagrass fairways and roughs at 2 to 4 wk intervals. For warm season grass sod production, apply 14.5 to 43.6 ounces per acre at 2 to 4 wk intervals. Use a spray volume of 1 to 2 gallons of water per 1000 sq ft. A nonionic surfactant may improve product performance. Do not irrigate for 4 hours after application or mow until 1 day after application.

Growth Regulators for Turfgrasses (continued)

Turfgrass	Brand	Amount of Formulation Per Acre	Pounds Active Ingredient Per Acre	Precautions and Remarks
Warm Season Grasses— Well-Maintained Turf: Foliar Suppression (continued)	trinexapac-ethyl (Primo Maxx) 1 MEC or (T-Nex) 1 AQ (Primo WSB) 25 WP	2.7 to 13 fl oz 1.35 to 6.5 oz	0.042 to 0.085 0.042 to 0.085	Application rates are for mowing heights of less than or equal to 0.5 inch Apply to actively growing, non-stressed turf. Rate varies with turfgrass species. See label for specific rate and other directions and precautions. Repeat applications can be made but do not exceed a total of 21.4 pints per acre per year of Primo Maxx or a total of 174 ounces per acre per year of Primo WSB. Do not exceed a total of 19 pints per acre per year of T-Nex. Refer to the respective Primo label for guidelines regarding mowing prior to and following applications. Mix with 0.5 to 4 gallons of water per 1,000 sq ft (20 to 174 gallons per acre). Primo can be applied to putting greens. See label for directions.
Warm Season Grasses—Low-Maintenance Turf: Seedhead and Foliar Suppression	glyphosate (Roundup Pro) 4 lb/gal	6 fl oz/10 to 25 gal water	0.2	Apply to bahiagrass only. Apply after full greenup of the bahiagrass (about late May) and make only one application per year. Do not apply to turf less than 3 years old. Treated turf may appear less dense and temporarily discolored. Optimum results may not be obtained if rainfall or overhead irrigation occurs within 6 hours following application. This is a nonselective herbicide. If application exceeds the above recommended rates, it can result in permanent loss of turf.
	(Touchdown Pro) 3 LC	0.375 to 4 pt/10 to 40 gal water	0.14 to 1.5	Touchdown Pro may be used on dormant or actively growing bermudagrass and bahiagrass turf described in "GENERAL USE AREAS" section of label. May be tank mixed with 0.25 to 2 ounces of Oust for residual weed control. Check label for correct rates. Touchdown Pro will control winter annual weeds less than 6 inches tall and also 4-to 6-leaf tall fescue in dormant turf. Use only on well-established bermudagrass. Injury may occur, but regrowth will occur under moist conditions. Bahiagrass vegetative growth and seedheads may be suppressed approximately 45 days when applied 1 to 2 wk after spring greenup and before seedhead emergence. A second application at 45 days will extend suppression to approximately 120 days.
	imazapic (Plateau) 2 ASU	2 fl oz	0.031	Only government entities may buy Plateau. Used for bahiagrass seedhead suppression. Apply to bahiagrass in spring after full greenup but approximately 3 to 4 weeks prior to expected seedhead emergence or 7 to 10 days after mowing. Do not apply to wetlands. Add a surfactant according to label directions. Bahiagrass may appear less dense and discolored following application.
	imazapic (Panoramic) 2 SL	2 to 3 fl oz	0.031	May be used for seedhead suppression of bahiagrass or tall fescue turf areas including industrial turf, golf courses, and non-residential areas. Apply 2-3 ounces/A for tall fescue seedhead suppression prior to seedhead emergence. Apply 2 ounces/acre after bahiagrass greenup but prior to seedhead emergence. Temporary turf discoloration may occur.
	imazapic + glyphosate (Journey) 2.25 AS	11 to 32 fl oz	0.19 to 0.56	Use in noncrop areas. Temporary turf discoloration may occur. Apply 4 to 8 fluid ounces per acre on a small area first to determine rate needed for desired results. Do not use with methylated seed oil. Do not apply to drought-stressed turf. Apply after full turf greenup.

Growth Regulators for Turfgrasses (continued)

Turfgrass	Brand	Amount of Formulation Per Acre	Pounds Active Ingredient Per Acre	Precautions and Remarks
Warm Season Grasses—Low-Maintenance Turf: Seedhead and Foliar Suppression (continued)	imazethapyr + imazapyr (Event) 1.46 lb/gal	8 to 10 fl oz	0.09 to 0.11	Apply to bahiagrass only. Apply after the turf is at 100% greenup and has at least 2 inches of vertical growth. The addition of a nonionic surfactant containing at least 80% active ingredient at 0.25% v/v of the spray (2 pints per 100 gallons of spray mixture) is required. Do not use on newly established stands less than 1 year old or on highly managed turf. Do not reseed before 3 months after application. See label for herbicide tank mix options. Follow label directions and precautions.
	maleic hydrazide (Retard) 2.5 lb/gal	1.3 gal/50 gal water	3	Apply to bahiagrass only. Apply in late spring but before seedheads appear. Applications made after seedhead appearance will suppress subsequent seedheads. Do not apply to turf less than 3 years old and do not reseed within 3 days after application. Treated turf may appear less dense and temporarily discolored. Optimum results may not be obtained if rainfall or overhead irrigation occurs within 12 hours following application. Remove excess grass clippings and leaves before application. Do not add a surfactant. Follow label directions and precautions. A repeat application may be needed 6 weeks after initial application.
	(Royal Slo-Gro) 1.5 lb/gal	2 gal/30 to 50 gal water	3	
	(Liquid Growth Retardant) 0.6 lb/gal	5 gal/45 gal water	3	
	mefluidide (Embark 2-S) 2 S	2 qt/15 to 150 gal water	1	Apply to bermudagrass only. Apply in late spring until about 2 weeks before seedhead appearance. Do not apply to turf within 4 growing months after seeding, and do not reseed within 3 days after application. Treated turf may appear less dense and temporarily discolored. Optimum results may not be obtained if rainfall or overhead irrigation occurs within 8 hours following application. Remove excess grass clippings and leaves before application. Adding 1 to 2 quarts of a nonionic surfactant per 100 gallons of spray solution may enhance suppression; however, discoloration may also be increased. Follow label directions and precautions.
	sulfometuron methyl (Oust) 75 DG	0.5 oz/30 to 50 gal water	0.02 lb	Apply to bahiagrass in late spring or early summer before seedheads appear. Do not apply to wetlands or where runoff water may flow onto agricultural lands or forests. Injury of desirable trees may result if applications are made near plants or where their roots extend or may be subjected to runoff from treated areas. Do not apply to turf less than 3 years old. Treated turf may appear less dense and temporarily discolored. Do not add a surfactant. Follow label directions and precautions.
	sulfometuron methyl + chlorsulfuron (Landmark MP) 50 + 25 DG	0.9 oz	0.042	For established bermudagrass and centipede-improved turf. Temporarily suppresses foliar and seedhead growth while controlling many grass and broadleaf weeds. Apply 30 days after breaking dormancy or either late fall or early winter. Landmark MP may discolor or cause top kill of desired turf species. Do not apply to turf less than 1 year old. Annual retreatments may reduce turf vigor.
	(Landmark II MP) 56.25 + 18.75 DG	1.0 oz	0.047	
	sulfometuron methyl + metsulfuron methyl (Oust Extra) 56.25 + 15 DG	0.5 to 2 oz	0.022 to 0.088	For use on well-established, unimproved bermudagrass and centipedegrass. Apply 30 days after breaking dormancy. Can also be applied in late fall or early winter depending on weed presence. Oust Extra can be tank mixed with 3 to 4 pounds active ingredient per acre MSMA on bermudagrass during the summer. Do not add a surfactant.

Growth Regulators for Turfgrasses (continued)

Turfgrass	Brand	Amount of Formulation Per Acre	Pounds Active Ingredient Per Acre	Precautions and Remarks
Annual Bluegrass: Suppression	flurprimidol (Cutless 50 W) 50 WP	0.25 to 0.5 lb/50 to 100 gal water	0.12 to 0.25	Apply to actively growing bentgrass putting greens in spring after third or fourth mowing or in the fall. Repeat, if necessary, at 3- to 6-week intervals, not to exceed 2 pounds per acre per growing season. Delay overseeding 2 weeks after application. Make final fall application 8 weeks before onset of winter dormancy.
		1 to 1.5 lb/50 to 200 gal water	0.5 to 0.75	Apply to bentgrass, Kentucky bluegrass, and perennial ryegrass in late spring-early summer and/or late summer-early fall. Time the second application to occur at least 3 months before expected winter dormancy. Management practices that encourage vigorous growth of perennial turfgrass following application will enhance conversion. *Poa annua* discoloration will be visible 7 to 10 days after treatment and last for 3 to 6 weeks. Do not apply to putting greens. Treated areas should receive 0.5 inch of irrigation within 24 hours after application. Resume mowing 3 to 5 days after application.
	flurprimidol + trinexapac-ethyl (Legacy) 1.51 SL	5 to 30 fl oz	0.059 to 0.354	Use in cool season turfgrasses, such as bentgrass greens and fairways, Kentucky bluegrass, and perennial ryegrass. Repeat applications at 2- to 6-week intervals. Annual bluegrass suppression is gradual and could take several growing seasons. Start treatments in early spring and continue through early fall.
	maleic hydrazide (Retard) 2.25 lb/gal	1 qt/30 to 40 gal water	0.56	Treat after two normal mowings but before seedhead appears. Applications made after seedhead appears will suppress subsequent seedheads. Do not apply to golf greens. Do not apply to turf less than 3 years old, and do not reseed within 3 days after application. Treated turf may appear less dense and temporarily discolored. Optimum results may not be obtained if rainfall or overhead irrigation occurs within 12 hours following application. Remove excess grass clippings and fallen leaves before application. Do not add a surfactant. Follow label directions and precautions for use on fairways.
	(Royal Slo-Gro) 1.5 lb/gal	2 qt/30 to 40 gal water	0.75	
	(Liquid Growth Retardant) 0.6 lb/gal	1.25 gal/30 to 40 gal water	0.75	
	mefluidide (Embark 2-S) 2 S	0.5 pt/15 to 150 gal water	0.125	Apply after uniform greenup but before first appearance of seedheads. Do not apply to turf within 4 growing months after seeding, and do not reseed within 3 days after application. Treated turf may appear less dense and temporarily discolored. Optimum results may not be obtained if rainfall or overhead irrigation occurs within 8 hours following application. Remove excess grass clippings and leaves before application. Adding 1 to 2 quart of a nonionic surfactant per 100 gallons of spray solution enhances suppression; however, discoloration may also be increased. Follow label directions and precautions for use of fairways and tees.
	(Embark) 0.2	2 to 5 pt/15 to 150 gal water	0.05 to 0.125	
	paclobutrazol (31-3-9 Fertilizer with TGR *Poa annua* Control 0.42%)	128 lb	0.5	Apply only to bentgrass, Kentucky bluegrass, perennial ryegrass fairways, or bentgrass greens with less than a 70% *poa annua* infestation. Follow label directions and precautions. Note: This product supplies 0.9 pound N per 1,000 sq ft.
	prohexadione calcium (Anuew) 27.5 WG	0.9 to 1.75 oz	0.015 to 0.03	Apply to overseeded hybrid bermudagrass during periods of active *poa annua* growth at 3 to 4 wk intervals.

Growth Regulators for Turfgrasses (continued)

Turfgrass	Brand	Amount of Formulation Per Acre	Pounds Active Ingredient Per Acre	Precautions and Remarks
Annual Bluegrass: Suppression (continued)	(15-0-29 High K Fertilizer with TGR *Poa annua* Control 0.34%)	98 lb to 146 lb	0.33 to 0.5	Apply only to bentgrass, zoysiagrass, Kentucky bluegrass, and Kentucky bluegrass/perennial ryegrass fairways, tees, and roughs, as well as bentgrass greens with less than 70% *poa annua* infestation. Note: This product supplies 0.5 pound N per 1,000 sq ft.
	(TGR Turf Enhancer 2 SC or Trimmit 2 SC) 2 SC	6.4 to 48 fl oz/43 to 200 gal water	0.1 to 0.75	Apply on hybrid bermudagrass, bentgrass, perennial ryegrass, and Kentucky bluegrass/perennial ryegrass fairways, tees, and roughs. Can also be applied to bentgrass putting greens. Apply in spring after greenup or regrowth has begun and after mowing once or twice. Apply with a nonburning fertilizer. Apply 0.25 inch of water within 24 hours after application to remove product from foliage and onto soil surface. See label for rates and other directions for applications to bentgrass putting greens and overseeded bermudagrass. Do not apply more than 3 times annually. Do not use on areas containing more than 70% *poa annua*. For bentgrass putting greens, do not apply more than 0.25 pound active ingredient per acre per application.
	ethephon (Proxy) 2 SL	1.7 gal	3.4	May be used to suppress annual bluegrass seedheads and growth of other cool season turfgrasses including golf course greens, fairways, tees, and roughs. Do not use an adjuvant. Do not apply to stressed turfgrass or where excessive thatch is present. Scalping may occur on bentgrass surfaces after application. Consult label for repeat application intervals.
Overseeded Bermudagrass Turf: Foliar Suppression	flurprimidol (Cutless 50 W) 50 WP	0.75 to 3 lb/50 to 200 gal water	0.37 to 1.5	Rates depend upon grass species and cultivar. Apply to zoysiagrass, Tifway, Tifgreen, and common bermudagrass in late spring-early summer and/or late summer-early fall. Time the second application to occur 8 to 10 weeks before expected winter dormancy. Do not apply to putting greens. Do not exceed 1.5 pound per acre per application on coarse-textured soils. Treated areas should receive 0.5 inch of irrigation within 24 hours after application. Resume mowing 3 to 5 days after application.
	flurprimidol + trinexapac-ethyl (Legacy) 1.51 SL	5 to 30 fl oz	0.059 to 0.354	Use in cool season turfgrasses, such as bentgrass greens and fairways, Kentucky bluegrass, and perennial ryegrass. Repeat applications at 2- to 6-week intervals. Annual bluegrass suppression is gradual and could take several growing seasons. Start treatments in early spring and continue through early fall.
	maleic hydrazide (Royal Slo-Gro) 1.5 lb/gal (Liquid Growth Retardant) 0.6 lb/gal	1.5 gal/50 gal water 3.3 gal/50 gal water	2.25	Apply in late September or early October to inhibit bermudagrass growth and allow winter overseeding to establish. Overseed no sooner than 48 hours after application. Follow label directions and precautions for use on greens and fairways.
	paclobutrazol (TGR Turf Enhancer 2 SC or Trimmit 2 SC) 2 SC	6.4 to 16 fl oz/43 to 200 gal water	0.1 to 0.25	Apply any time after overseeded turf has successfully established itself. Do not apply after March 15 to avoid delay in bermudagrass green-up. Apply with 0.25 to 0.5 pound N per 1,000 sq ft of a nonburning fertilizer. Apply 0.25 inch of water within 24 hours after application to remove product from foliage and onto soil surface. Repeat applications can be made but *do not apply* more than 3 times annually. Do not use on areas containing more than 70% *poa annua*. Do not seed within 6 weeks prior to or 2 weeks after application. Do not apply to 'Tifdwarf' putting greens.

Growth Regulators for Turfgrasses (continued)

Turfgrass	Brand	Amount of Formulation Per Acre	Pounds Active Ingredient Per Acre	Precautions and Remarks
Overseeded Bermudagrass Turf: Foliar Suppression (continued)	prohexadione calcium (Anuew) 27.5 WG	0.9 to 1.75 oz	0.015 to 0.03	To enhance overseeding establishment, apply to hybrid bermudagrass 3 to 5 days prior to seeding. Delay verticutting, spiking or scalping for 1 to 2 days after application.
	trinexapac-ethyl (Governor) 0.17 G	129 to 165 lb	0.22 to 0.28	Apply before verticutting, scalping, or spiking the bermudagrass. Apply 1 to 5 days before overseeding. To minimize yellowing, use iron at recommended rates or available nitrogen at 0.2 to 0.5 pound per 1,000 square feet.
	(Primo Maxx) 1 MEC or (T-Nex) 1 AQ (Primo WSB) 25 WP	6 to 44 fl oz 2.75 to 21.8 oz	0.08 to 0.34 0.08 to 0.34	Application rate varies with turfgrass species and height of cut. Apply to actively growing, nonstressed turf. More growth suppression occurs at lower mowing heights. See label for specific rate and other directions and precautions. Repeat applications can be made but do not exceed a total of 21.4 pints per acre per year of Primo Maxx or a total of 174 ounces per acre per year of Primo WSB. Do not exceed 19 pints per acre per year of T-Nex. Refer to the respective Primo label for guidelines regarding mowing prior to and following application. Mix with 0.5 to 4 gallons of water per 1,000 sq ft (20 to 174 gallons per acre). Primo can be applied to putting greens. See label for directions.
Lawn Edging	maleic hydrazide (Retard) 2.25 lb/gal	1.33 gal/100 gal water	3	Apply in spring to a 6-inch band along sidewalks. Consult instructions on applicator for delivery dosage.
	(Royal Slo-Gro) 1.5 lb/gal	2 gal/100 gal water	3	
	(Liquid Growth Retardant) 0.6 lb/gal	6.67 gal/100 gal water	4	
	mefluidide (Embark) 0.2	1.36 gal/174 gal water	0.27	For Kentucky bluegrass, tall fescue, chewings fescue, red fescue, perennial ryegrass, and St. Augustinegrass. For bermudagrass, use 5.45 gallons in 174 gallons water. Apply in 6- to 12-inch bands. Avoid overlapping.
	trinexapac-ethyl (Governor) 0.17 G	100 to 259 lb	0.17 to 0.44	Do not exceed 2.5 pounds active ingredient per acre per year. These rates should provide 50% turf growth suppression for 4 weeks with minimal yellowing.
	(Primo Maxx) 1 MEC (T-Nex) 1 AQ (Primo WSB) 25 WP			Apply 0.75 to 2 ounces per 1,000 linear feet of Primo Maxx or T-Nex, or 0.4 to 2 ounces per 1,000 linear ft of Primo WSB. Apply to actively growing, nonstressed turf. Apply along perimeter of lawns, sidewalks, curbs, parking lots, driveways, flower beds, or fences. Apply in an 8- to 12-inch band along the perimeter of the lawn to reduce growth of turf into adjacent areas. Application rate varies with turf species. Follow label directions for repeat applications and other precautions.

Aquatic Weed Control

R. J. Richardson, Crop and Soil Sciences Department, and K. D. Getsinger, U.S. Army Engineer Research and Development Center, Vicksburg, MS, and Adjunct Professor, Crop and Soil Sciences Department, NC State University

Several options, including hand removal, cultural, mechanical, biological, and chemical control techniques are available for the management of aquatic weeds. The applicator should choose the most efficacious, environmentally acceptable, and cost-effective alternative that is available for a particular weed problem. The site-specific management strategy to use in a given situation will depend on the intended use of the body of water, fish, and wildlife populations that may be impacted, type of environment in which the weed problem occurs, and the particular weed species of concern. Before selecting your management strategy, be sure to have the weed(s) of concern identified by a qualified individual.

Assistance in weed identification is available from the Cooperative Extension center in your county. Additional information on management techniques also may be obtained from the local Extension center; ask for AG437, Weed Management in Small Ponds; AG-438, Weed Control in Irrigation Water Supplies; and AG-449, Hydrilla: A Rapidly Spreading Aquatic Weed in North Carolina. Information on pond construction, stocking, and general pond management may be found in AG-424, Pond Management Guide. Additional information may be found on the Aquatic Weed Management Web site: go.ncsu.edu/aquatic-weed-management.

For the purpose of description and management, aquatic weeds may be grouped either on the basis of their botanical relationships or on the basis of their growth habits. Most plants in each group are managed similarly, with some exceptions.

Aquatic Plant Groups — Grouping of Aquatic Plants on the Basis of Botanical Relationships

Category and Description	Examples

1. *Algae* — These plants may be either microscopic or visible to the naked eye, exist as single cells or occur in clusters or filaments containing many cells and may be either free floating (planktonic) or attached to the soil, rocks, or vegetation. Filamentous algae may be unbranched, slightly or highly branched, or net-like. Some planktonic algae are mobile. Certain types of algae (macroalgae) may be large, very coarse, and resemble submersed vascular plants. Most algae (except macroalgae) usually require magnification to be identified accurately. Algae do not contain vascular (water conducting) tissues, consequently all chemicals used for algae control have only contact activity. Algae reproduce by cell division, fragmentation, and sexually by spores.	**Filamentous Algae**
	Bluegreens or
	  Cyanobacteria
	Giant *Lyngbya*
	Green algae
	  Oedogonium
	  Hydrodictyon (water net)
	  Spirogyra
	  Pithophora
	Planktonic Algae
	Bluegreens or
	  Cyanobacteria
	  Lyngbya
	  Anabaena
	  Oscillatoria
	  Microcystis
	Euglenoids (*Euglena*)
	Macroalgae
	Muskgrass (*Chara*)
	Stonewort (*Nitella*)

2. *Mosses* — These plants are visible to the naked eye and resemble delicate, leafy submersed plants. The mosses lack vascular tissues or roots, but usually are attached to the soil. Mosses reproduce sexually by spore production.	*Fontinalis* *Sphagnum* (peat moss)
3. *Ferns* — These plants are visible to the naked eye, either free floating or rooted to the bottom, occasionally forming loosely consolidated floating mats. Ferns have vascular tissues and reproduce by vegetative propagation and sexually by spores.	Giant salvinia (*Salvinia molesta*) Mosquito fern (*Azolla* spp.) Water clover (*Marsilea quadrifolia*) Water spangles (*Salvinia minima*)

Aquatic Plant Groups — Grouping of Aquatic Plants on the Basis of Botanical Relationships (continued)

Category and Description	Examples
4. *Vascular flowering plants* — These plants may be rooted or unrooted, free floating, submersed, floating-leaved, or emergent. Most reproduce vegetatively by means of rhizomes, stolons, and various other vegetative perennating structures including turions and tubers. Most also produce flowers and may set seeds. This group has a vascular system that shows varying degrees of development from rudimentary in the case of the duckweeds and submersed species to very complex and highly developed in emergent plants and includes annual and perennial herbaceous forms and several woody species.	Bald cypress (*Taxodium distichum*) Bladderwort (*Utricularia* spp.) Bulrushes (*Scirpus* spp.) Cattail (*Typha* spp.) Duckweed (*Lemna* spp. and *Spirodela* spp.) Hydrilla *(Hydrilla verticillata)* Naiads (*Najas* spp.) Pondweeds (*Potamogeton* spp.) Rushes (*Juncus* spp.) Spikerushes (*Eleocharis* spp.) Waterhyacinth (*Eichhornia crassipes*) Watermilfoils (*Myriophyllum* spp.)
1. *Submersed plants* — Plants in this group grow beneath the surface of the water and may be rooted to the bottom or free floating, with or without roots. Flowers usually are produced above the surface of the water and occasionally may be supported by specialized floatation structures. Some species will produce emergent floral spikes that extend several inches above the surface of the water and are covered with bracts that resemble leaves. Submersed plants usually have poorly developed vascular systems and very limited structural tissue and depend on the water's buoyancy for support. Filamentous algae and macroalgae also could be considered submersed plants.	American elodea (*Elodea canadensis* and *E. nuttallii*) Bladderwort (*Utricularia* spp.) Brazilian elodea (*Egeria densa*) Brittle naiad (*Najas minor*) Coontail (*Ceratophyllum demersum*) Creeping rush (*Juncus repens*) Eurasian watermilfoil (*Myriophyllum spicatum*) Fanwort (*Cabomba caroliniana*) Hydrilla (*Hydrilla verticillata*) Parrotfeather (*Myriophyllum aquaticum*) Pondweeds (*Potamogeton* spp.) Proliferating spikerush (*Eleocharis baldwinii*) Southern naiad (*Najas guadalupensis*) Variable-leaf milfoil (*Myriophyllum heterophyllum*) Widgeongrass (*Ruppia maritima*) Wild celery (*Vallisneria americana*)

Aquatic Weed Groups — Grouping of Aquatic Plants on the Basis of Growth Habit
NOTE: Some species have growth habits that overlap and may be listed more than once.

Category and Description	Examples
1. *Submersed plants* — Plants in this group grow beneath the surface of the water and may be rooted to the bottom or free floating, with or without roots. Flowers usually are produced above the surface of the water and occasionally may be supported by specialized floatation structures. Some species will produce emergent floral spikes that extend several inches above the surface of the water and are covered with bracts that resemble leaves. Submersed plants usually have poorly developed vascular systems and very limited structural tissue and depend on the water's buoyancy for support. Filamentous algae and macroalgae also could be considered submersed plants.	American elodea (*Elodea canadensis* and *E. nuttallii*) Bladderwort (*Utricularia* spp.) Brazilian elodea (*Egeria densa*) Brittle naiad (*Najas minor*) Coontail (*Ceratophyllum demersum*) Creeping rush (*Juncus repens*) Eurasian watermilfoil (*Myriophyllum spicatum*) Fanwort (*Cabomba caroliniana*) Hydrilla (*Hydrilla verticillata*) Parrotfeather (*Myriophyllum aquaticum*) Pondweeds (*Potamogeton* spp.) Proliferating spikerush (*Eleocharis baldwinii*) Southern naiad (*Najas guadalupensis*) Variable-leaf milfoil (*Myriophyllum heterophyllum*) Widgeongrass (*Ruppia maritima*) Wild celery (*Vallisneria americana*)
2. *Free-floating plants* — Plants in this group float on the surface of the water and may lie flat on the water or be raised well above the surface. These plants, with the exception of the duckweeds, watermeal, and mosquito ferns, have well-developed vascular systems and substantial supportive tissues. Most form true roots. Flowers extend above the surface of the water in the flowering plants.	Duckweeds (*Lemna* spp. and *Spirodela* spp.) Floating heart (*Nymphoides aquatica*) Frogbit (*Limnobium spongia*) Giant salvinia (*Salvinia molesta*) Mosquito fern (*Azolla caroliniana*) Waterhyacinth (*Eichhornia crassipes*) Waterlettuce (*Pistia stratiotes*) Watermeal (*Wolffia* spp.)

Aquatic Weed Groups — Grouping of Aquatic Plants on the Basis of Growth Habit (continued)

Category and Description	Examples
3. *Floating leaf plants* — These plants are rooted in the bottom and have their leaves attached to long, tough stems that extend to the surface from depths up to 6 ft or more. The leaves float directly on the surface of the water. Mature leaves of some species may push well above the surface into an emergent position. Most of these plants have extensive root and rhizome systems and well-developed vascular systems and supportive tissues. Flowers float just above the surface or are extended well above the surface on a tough stem. A few nonvascular representatives.	American lotus (*Nelumbo lutea*) Fragrant waterlily (*Nymphaea odorata*) Illinois pondweed (*Potamogeton illinoiensis*) Spatterdock (*Nuphar luteum*) Water clover (*Marsilea quadrifolia*) Watershield (*Brasenia schreberi*)
4. *Emergent plants* — These plants grow rooted in the bottom with their leaves and green stems extending well above the surface of the water. A few species also may form floating mats. All have extensive root and rhizome systems and well-developed vascular systems and supportive tissues. Reproduction occurs vegetatively by rhizomes and stolons; floating mat-forming species also reproduce readily by stem fragmentation. Most flower prolifically and form many seeds.	**Broadleaf Species** Arrow arum (*Peltandra virginica*) Arrowhead (*Sagittaria* spp.) Asian spiderwort (*Murdannia keisak*) Frogbit (*Limnobium spongia*) Lizard's tail (*Saururus cernuus*) Pickerelweed (*Pontederia cordata*) Smartweeds (*Polygonum* spp.) **Mat-forming Broadleaf Species** Alligatorweed (*Alternanthera philoxeroides*) Creeping waterprimrose (*Ludwigia hexapetala*) Water pennywort (*Hydrocotyle* spp.) Water willow (*Justicia americana*) **Sedges, Rushes, Spikerushes, and Grasses** Bulrush (*Scirpus* spp.) Cattail (*Typha* spp.) Common reed (*Phragmites australis*) Flat sedge (*Carex* spp.) Foursquare (*Eleocharis quadrangulata*) Maidencane (*Panicum hemitomon*) Rushes (*Juncus* spp.) Sedge (*Cyperus* spp.) Soft rush (*Juncus effusus*) Softstem bulrush (*Scirpus validus*) Southern wildrice (*Zizaniopsis miliacea*) Spikerushes (*Eleocharis* spp.) Threesquare bulrush (*Scirpus americanus*) Torpedograss (*Panicum repens*) Water paspalum (*Paspalum repens*) Woolgrass (*Scirpus cyperinus*) **Other Common Species** Bur-reed (*Sparganium americanum*) Scouring rush (*Equisetum hymale*)
5. *Woody plants* — These are obligate, aquatic species of trees usually growing totally flooded or in saturated soils, but occasionally occur in upland areas (usually planted there). Some form systems of "knees" to provide aeration for the root systems. They are deciduous, dropping leaves in the autumn, and are rarely if ever vegetative during winter months.	Bald cypress (*Taxodium distichum*) Pond cypress (*Taxodium ascendens*) Tupelo (*Nyssa aquatica*)

Biological Control of Aquatic Weeds with Triploid Grass Carp

While the triploid, sterile grass carp is a cost-effective control method, it is best suited for use in small ponds, where submersed aquatic plants are not required for fish and wildlife habitat. Grass carp are effective on most submersed weeds. They generally are less effective on algae and weeds in the floating and emergent groups. Refer to the chart below for information on the relative effectiveness of grass carp for different weeds.

Grass carp are normally stocked at 15 fish per acre in small ponds. In larger ponds, they are usually stocked at 15 to 20 fish per vegetated acre. Large fish (minimum of 8 to 10 inches long) should be stocked to prevent loss due to predation by large bass and wading birds. If the surface of the pond is completely covered with vegetation, some limited herbicide application or mechanical removal of weeds from a portion of the pond will be necessary before stocking to allow oxygen to reach the underlying water. Grass carp may be stocked at any time of the growing season, but best results are usually obtained by a late summer or fall stocking.

No permit is required to purchase up to 150 triploid grass carp for stocking a private pond. At a stocking rate of 15 fish per acre of water, 150 triploid grass carp are adequate to control vegetation in a 10-acre pond. A permit from the Wildlife Resources Commission is required for larger stockings. Grass carp may be purchased from a licensed distributor. Permits, a list of certified distributors, and additional information on stocking of triploid grass carp may be obtained from the Wildlife Resources Commission, Chief of Inland Fisheries, 1721 Mail Service Center, Raleigh, NC 27699-1721, or call at (919) 707-0220.

Biological Control of Aquatic Weeds with Triploid Grass Carp

Weed	Relative Effectiveness	Comments
Algae Filamentous (green and bluegreen) and planktonic	Poor	High stocking rates (60 to 75 or more fish per acre) with small fish (4 to 6 inches size) are required to achieve temporary control; control usually decreases as fish grow larger and are unable to feed on the algae.
Macroalgae Chara and Nitella	Good to Excellent	Chara usually is beneficial to fish and wildlife.
Floating and Floating-Leaved Weeds Duckweeds, watermeal	Poor	Small fish at very high stocking rates (see filamentous algae above) may give control; larger fish at normal stocking rates usually are not effective.
Water ferns (Azolla and Salvinia)	Fair to Poor	
Alligatorweed, water lilies, water primrose, lotus, watershield, spadderdock, waterhyacinth	Poor	Grass carp may feed lightly on weeds in this group, but control is usually unacceptable.
Emergent and Marginal Weeds Cattails, rushes, common reed, bulrushes, pickerelweed, pennywort, arrowhead	Poor	Grass carp may feed lightly on weeds in this group, but control is usually unacceptable.
Submersed Weeds	Good to Excellent	Most rooted and free-floating submersed weeds in ponds are readily controlled with triploid grass carp; control may be poorer on the watermilfoils, particularly Eurasian waterfoil.

Chemical Control of Aquatic Plants

Plant	Herbicide, Formulation, and Mode of Action Code[1]	Amount of Formulation	Active Ingredient Rate or Concentration	Precautions and Remarks[2]
Algae, filamentous and planktonic	copper complex (various)	0.6 gal/acre ft	0.2 ppm	Dilute with water in ratio of at least 9-to-1 and apply uniformly. For best results, apply on a clear day and break up floating mats of filamentous algae before treatment. **Warning: Copper is toxic to fish.**
	copper sulfate (various)	See label	0.5 to 1 ppm	Apply crystals or powder at early stage of growth by any method to give rapid and uniform dispersion. For best results, apply on a clear day. Do not apply to muddy water. **Warning: Copper is toxic to fish.** Copper products formulated with a chelating agent (copper complex) have a greater margin of safety to fish.
	diquat (Reward) 2 lb/gal MOA 22	See label	0.18 to 0.37 ppm	For certain filamentous algae—*Pithophora* spp. and *Spirogyra* spp. Check label for application instructions. For best results, break up floating mats before treatment.
	sodium carbonate peroxyhydrate (various)	See label	0.3 to 1.7 ppm	Apply with 8 to 10 hours of daylight remaining. Do not reapply within 48 hours.
Algae, macro, chara, nitella	copper complex (Cutrine-Plus Granular) 3.7 G (Cutrine-Plus) 0.9 lb/gal (K-Tea) 0.8 lb/gal	60 lb/surface acre 1.2 gal/acre ft 1.7 to 3.4 gal/acre ft	2.2 lb/acre 0.4 ppm 0.5 to 1.0 ppm	Distribute granular formulation evenly over infested area when plants are young. If chara is in water less than 3 ft deep or growth is near the surface, the liquid formulation may be used. Dilute with water in ratio of at least 9-to-1 and apply uniformly. **Warning: Copper is toxic to fish.**
Algae, Pithophora and cladophora	flumioxazin (Clipper) 51% MOA 14	6 to 12 oz/A	3 to 6 ai/A or 100 to 400 ppb	Early morning applications may be more effective when water pH tends to be lower. If vegetation is dense, treat in sections to avoid reducing dissolved oxygen. Water pH greater than 7.5 will reduce effectiveness.
Floating Weeds (except watermeal)	2,4-D amine (various) MOA 4	See label	2 to 4 lb/acre	Thorough wetting of foliage is essential. Apply with 100 gallons of water per acre. Use low pressure, large nozzle, and spray thickener.
	bispyribac (Tradewind) 80% MOA 2	1 to 2 oz/A	0.8 to 1.6 oz ai/A	Controls duckweed, mosquito fern, salvinia, water hyacinth, water lettuce, and water pennywort. Apply with at least 30 gpa water volume. Include aquatic-approved adjuvant.
	carfentrazone (Stingray) 1.9 lb/gal, MOA 14	3.4 to 13.5 fl oz/acre	0.05 to 0.2 lb/acre	Controls water lettuce, waterhyacinth, salvia, duckweed, mosquito fern, and water spinach. Rates vary according to target species. Methylated seed oil or nonionic surfactant (aquatic-approved) recommended.
	diquat (Reward) 2 lb/gal MOA 22	0.5 to 0.75 gal/surface acre	1 to 1.5 lb/acre	Weeds controlled: pennywort, salvinia, waterhyacinth, waterlettuce. Apply in a spray volume of 150 to 200 gallons of water per acre plus aquatic-approved nonionic surfactant.
		1 gal/surface acre	2 lb/acre	For duckweed control, apply in a spray volume of 50 to 150 gallons of water per acre. Take care to cover all plants on water and damp marginal areas. Will require retreatment. An aquatic-approved nonionic surfactant at 0.5% by volume may be used.
	florpyrauxifen (Procellacor)	1 to 2 prescription dose units (PDU)	1 to 2 prescription dose units (PDU)	Controls several emergent species including alligatorweed. Labelled rates provided as PDUs. Product only available to approved aquatic applicators. Addition of aquatic-approved nonionic surfactant is recommended.
	glyphosate (various) MOA 9	See label	See label	For control of waterlilies, spadderdock, and lotus, apply as foliar spray on a calm day when there is little to no wave action. Vegetation must be on or above the surface for treatment to be effective. A nonionic surfactant approved for aquatic use is required with some formulations. If applying from a boat, take care not to create waves that may wash the herbicide off floating leaves. Will not control small floating plants, such as azolla, duckweed, or watermeal.
	imazamox (Clearcast) 1 lb/gal, MOA 2	32 to 64 fl oz/acre	0.25 to 0.5 lb ai/acre 50 to 150 ppb	See label for specific weeds controlled. An aquatic-approved nonionic surfactant or methylated seed oil is recommended for foliar applications. Spot treatments may be made with up to 5% solution by volume.

Chemical Control of Aquatic Plants (continued)

Plant	Herbicide, Formulation, and Mode of Action Code[1]	Amount of Formulation	Active Ingredient Rate or Concentration	Precautions and Remarks[2]
Floating Weeds (except watermeal) (continued)	imazapyr (Habitat) MOA 2	1 to 4 pt/acre	0.25 to 1.5 lb/acre	Rates vary according to target species. Retreatment of some plants may be required. An aquatic-approved nonionic surfactant is recommended. Will not control small floating plants, such as azolla, duckweed, or watermeal.
	penoxsulam (Galleon) 2 lb/gal, MOA 2	2 to 5.6 fl oz/acre	0.03 to 0.09 lb/acre 5 to 150 ppb	An aquatic-approved nonionic surfactant is recommended for foliar applications.
	triclopyr (Renovate 3) MOA 4	0.5 to 2 gal/acre	1.5 to 6 lb/acre	Rates vary according to target species. Addition of aquatic-approved nonionic surfactant is recommended.
	topramezone (Oasis) 29.7%	up to 16 fl oz/acre	up to 0.35 lb/acre	Use of an aquatic-approved surfactant is recommended for foliar applications. Check label for specific irrigation restrictions.
	flumioxazin (Clipper) 51% MOA 14	6 to 12 oz/acre	3 to 6 ai/A or 100 to 400 ppb	Early morning applications may be more effective when water pH tends to be lower. If vegetation is dense, treat in sections to avoid reducing dissolved oxygen. Water pH greater than 7.5 will reduce effectiveness. A follow-up application will likely be needed for long-term watermeal control. Application with diquat, flumioxazin, or fluridone may provide enhanced watermeal control.
	fluridone (Sonar) 4 AS MOA 12	Ponds: 0.16 to 1.5 qt/acre	0.16 to 1 lb/acre 45 to 90 ppb	Product amount will depend on average depth of water body. Do not apply when there is substantial outflow from the pond. Do not apply as a spot treatment. See label for specific weeds controlled. For watermeal, use 45 to 90 ppb. Other floating species may be controlled with lower rates. Do not use treated water for irrigation for 7 to 30 days. See label for irrigation precautions. **Warning: 30 days may be insufficient restriction if pond water will be used to irrigate very sensitive crops, such as tobacco, tomatoes, or peppers.**
Emergent, Marginal, and Ditchbank Weeds	2,4-D amine (various) MOA 4	See label	2 to 4 lb/acre	Thorough wetting of foliage is essential. Apply in 100 to 400 gallons of water per acre. Use low pressure, large nozzle and spray thickener. An aquatic-approved adjuvant may improve efficacy.
	2,4-D granular (Navigate) 20 G (2,4-D Gran 20) 20 G MOA 4	150 to 200 lb/surface acre	30 to 40 lb/acre	Weeds controlled: arrowhead, bulrush, creeping waterprimrose, pickerelweed, smartweed, spadderdock, waterchestnut, waterlily, watershield. Rate depends upon species and depth of water. Check label. Apply early, when weeds are actively growing, with a rotary seeder. Spadderdock may require retreatment.
	bispyribac (Tradewind) 80% MOA 2	1 to 2 oz/A	0.8 to 1.6 oz ai/A	Controls alligatorweed and parrotfeather. Apply with at least 30 gpa water volume. Include aquatic-approved adjuvant.
	carfentrazone (Stingray) 1.9 lb/gal, MOA 14	6.7 to 13.5 fl oz/acre	0.2 lb/acre	Suppresses alligatorweed and waterprimrose. Methylated seed oil or nonionic surfactant (aquatic-approved) recommended.
	diquat (Reward) 2 lb/gal (Weedtrine) 0.4 lb/gal MOA 22	1 gal/surface acre	2 lb/acre	For control of cattails in ponds or lakes. For top kill, apply in 100 gal of water per acre with 0.25% to 0.5% nonionic surfactant. Apply before flowering for best results. Retreat as needed.
	florpyrauxifen (Procellacor)	1 to 2 prescription dose units (PDU)	1 to 2 prescription dose units (PDU)	Controls several emergent species including alligatorweed. Labelled rates provided as PDUs. Product only available to approved aquatic applicators. Addition of aquatic-approved nonionic surfactant is recommended.
	flumioxazin (Clipper) 51% MOA 14	6 to 12 oz/A	3 to 6 ai/A or 100 to 400 ppb	Early morning applications may be more effective when water pH tends to be lowest. If vegetation is dense, treat in sections to avoid reducing dissolved oxygen. Ensure adequate coverage of dense vegetation or a follow-up application may be necessary. Addition of aquatic-approved nonionic surfactant is recommended.

Chemical Control of Aquatic Plants (continued)

Plant	Herbicide, Formulation, and Mode of Action Code[1]	Amount of Formulation	Active Ingredient Rate or Concentration	Precautions and Remarks[2]
Emergent, Marginal, and Ditchbank Weeds (continued)	glyphosate (various) MOA 9	See label	See label	Rates vary according to target species. Retreatment of alligatorweed is necessary. Aquatic-approved nonionic surfactant is recommended. Note: The use of very hard water or water containing high concentrations of iron to prepare spray solutions may result in reduced efficacy of glyphosate.
	imazamox (Clearcast) 1 lb/gal MOA 2	32 to 64 fl oz/ acre	0.25 to 0.5 lb ai/ acre 50 to 500 ppb	See label for specific weeds controlled. An aquatic-approved nonionic surfactant or methylated seed oil is recommended for foliar applications. Spot treatments may be made with up to 5% solution by volume. Rates vary according to target species. Retreatment of some plants may be required. An aquatic-approved nonionic surfactant is recommended.
	imazapyr (Habitat) MOA 2	1 to 6 pt/acre	0.25 to 1.5 lb/ acre	Rates vary according to target species. Retreatment of some plants may be required. An aquatic-approved nonionic surfactant is recommended.
	penoxsulam (Galleon) 2 lb/gal, MOA 2	2 to 5.6 fl oz/ acre	0.03 to 0.09 lb/ acre 5 to 500 ppb	See label for specific weeds controlled and application details. An aquatic-approved nonionic surfactant is recommended.
	triclopyr (Renovate 3) MOA 4	0.5 to 2 gal/ acre	1.5 to 6 lb/acre	Rates vary according to target species. Addition of an aquatic-approved nonionic surfactant is recommended.
	topramezone (Oasis) 29.7%	up to 16 fl oz/ acre	up to 0.35 lb/ acre	Use of an aquatic-approved surfactant is recommended for all foliar applications. Check label for specific irrigation restrictions.
Submersed Weeds[3]	2,4-D granular (Navigate) 20 G, MOA 4	100 to 200 lb/ surface acre	20 to 40 lb/acre	Controls milfoils and certain other submersed species. Rate depends upon weed to be controlled and depth of water. Check labels for species and rates. Apply uniformly with a rotary spreader.
	bispyribac (Tradewind) 80% MOA 2	See label	10 to 45 ppb	Controls hydrilla, sago pondweed, and Eurasian watermilfoil. Do not apply in areas of high water flow or water diffusion. Refer to label for specific details on application rate based on water volume.
	carfentrazone (Stingray) 1.9 lb/gal MOA 4	0.286 to 5.75 gal/acre	200 ppb	Controls Eurasian watermilfoil. Apply in spring or early summer as a subsurface application or with an appropriate adjuvant to ensure sinking and mixing of the spray mix. Early morning applications may be more effective when water pH tends to be lowest. Water pH greater than 7.5 will reduce effectiveness.
	diquat (Reward) 2 lb/gal MOA 22	1 to 2 gal/ surface acre	2 to 4 lb/acre	Weeds controlled: bladderwort, coontail, elodea, naiads, pond weeds. Apply early in season by pouring directly into water in strips 40 feet apart. Later in season, as weeds reach surface, pour in strips 20 feet apart or inject a dilute solution. Not effective in turbid or muddy water. If vegetation is dense, treat in sections to avoid reducing dissolved oxygen.
	endothall (Aquathol K) 4.2 lb/gal (Aquathol Super K) 63 G	0.3 to 2.6 gal/ acre ft 2.2 to 17.6 lb/ acre ft	0.5 to 5 ppm	Weeds controlled: bass weed, bur reed, coontail, hydrilla (Aquathol K only), pondweeds, watermilfoil, water star grass. Rate depends upon weed species and type of treatment. Spot or marginal treatments require higher rates. Aquathol Granular is especially useful for spot or marginal treatments.
	florpyrauxifen (Procellacor)	1 to 5 prescription dose units (PDU)	1 to 5 prescription dose units (PDU)	Controls several submersed species including hydrilla and milfoils. Labelled rates provided as PDUs. Product only available to approved aquatic applicators.
	flumioxazin (Clipper) 51% MOA 14	See label	100 to 400 ppb	Early morning applications may be more effective when water pH tends to be lowest. If vegetation is dense, treat in sections to avoid reducing dissolved oxygen. Water pH greater than 7.5 will reduce effectiveness.

Chemical Control of Aquatic Plants (continued)

Plant	Herbicide, Formulation, and Mode of Action Code[1]	Amount of Formulation	Active Ingredient Rate or Concentration	Precautions and Remarks[2]
Submersed Weeds[3] (continued)	fluridone (Sonar) AS MOA 12	Ponds: 0.16 to 1 qt/acre Lakes: 0.2 to 4 qt/acre Canals: 2 qt/acre	0.16 to 1 lb/acre 0.2 to 4 lb/acre 2 lb/acre	Do not use water for irrigation for 7 to 30 days. See label for specific irrigation precautions. Application to canals should be made only if water flow can be restricted. **Warning: 30 days may be insufficient restriction if applied to small ponds and pond water will be used to irrigate very sensitive crops, such as tobacco, tomatoes, or peppers.**
	(Sonar SRP) MOA 12	Ponds: 3.2 to 30 lb/acre Lakes: 4 to 80 lb/acre Canals: 40 lb/acre Rivers: 40 lb/acre	0.16 to1.5 lb/acre 0.2 to 4 lb/acre 2 lb/acre 2 lb/acre	
	imazamox (Clearcast) 1 lb/gal, MOA 2	See label	50 to 500 ppb	Rates vary according to target species and depth to be treated. See label for specific weeds controlled and application details.
	penoxsulam (Galleon) 2 lb/gal, MOA 2	See label	5 to 150 ppb	Rates vary according to target species and depth to be treated. See label for specific weeds controlled and application details.
	triclopyr (Renovate 3 or OTF), MOA 4	See label	1.5 to 6 lb/acre 0.5 to 2.5 ppm	Controls milfoils and certain other submersed species. Rates vary according to target species and depth to be treated. See label for specific weeds controlled and application details.
	topramezone (Oasis) 29.7%	up to 16 fl oz/acre	up to 0.35 lb/acre	Rates vary according to target species and depth to be treated. See label for specific weeds controlled and application details. Check label for specific irrigation restrictions.

[1] Mode of Action (MOA) code developed by the Weed Science Society of America. Cooper compounds, endothall. and sodium carbonate peroxyhydrate have not been assigned codes.

[2] Also see comments for specific herbicides under "Table 7-25. Labeled Sites and Restrictions."

[3] Grass carp give cost-effective control on the majority of the weeds in this group and should be given consideration *before* using herbicides. See text at beginning of this section under *Biological Control of Aquatic Weeds with Triploid Grass Carp*. A permit is required to purchase more than 150 grass carp or for stocking in impoundments larger than 10 acres. Grass carp usually are **not effective** on filamentous algae, duckweed, watermeal, or any of the plants in the floating or emergent groups.

Waiting Period (in Days) Before Using Water After Application of Herbicides for Aquatic Weed Control

Herbicide	Irrigation[1]	Fish Consumption	Watering Livestock	Swimming
2,4-D (various formulations and manufacturers)	Water use restrictions vary by formulation and manufacturer. In general, if water is used for irrigating sensitive crops, 2,4-D should not be used. Turfgrasses are generally tolerant to low concentrations of 2,4-D. Also, many 2,4-D formulations are NOT labelled for aquatic use. Read the label before purchasing and/or use.			
Bispyribac (Tradewind)	Do not irrigate until concentrations are < 1 ppb	No restrictions	Do not water livestock until concentrations are ≤ 1 ppb	No restrictions
carfentrazone (Stingray)	1 to 14[2]	No restrictions	0 to 1	No restrictions
copper (Copper sulfate pentahydrate, including Bluestone and EarthTec; and complexed copper formulations, including Algae-Pro, Captain, Clearigate, Cutrine-Plus, Cutrine-Plus Granular, K-Tea, Komeen, etc.)	No restrictions	No restrictions	No restrictions	No restrictions
diquat (Reward)	3 to 5[3]	No restrictions	1	No restrictions
endothall (Aquathol K) (Aquathol Super K) (Hydrothol 191) (Hydrothol 191 granular)	No restrictions for many situations. See label for specific restrictions	No restrictions	7 to 25	No restrictions
Florpyrauxifen (Procellacor)	Do not use treated water for irrigation unless allowed by product label	No restrictions	Do not allow livestock to drink treated water unless allowed by product label	No restrictions
Flumioxazin (Clipper)	0 to 5[3]	No restrictions	No restrictions	No restrictions
fluridone (Sonar 4AS, Sonar SRP)	7 to 30[3]	No restrictions	No restrictions	No restrictions
Glyphosate (AquaMaster, Aqua Neat, Rodeo, Touchdown Pro)	No restrictions	No restrictions	No restrictions	No restrictions
imazamox (Clearcast)	0+[3]	No restrictions	No restrictions	No restrictions
Imazapyr (Habitat)	120	No restrictions	No restrictions	No restrictions
penoxsulam (Galleon)	Do not irrigate food crops until residues ≤ 1 ppb	No restrictions	No restrictions	No restrictions
sodium carbonate peroxyhydrate (GreenClean Pro, Pak 27)	No restrictions	No restrictions	No restrictions	No restrictions
topramezone (Oasis)	See label for specific irrigation restrictions	No restrictions	No restrictions	No restrictions
triclopyr (Renovate 3, Renovate OTF)	120 0 to established grass	No restrictions	Next growing season for lactating dairy animals	No restrictions

[1]Irrigation restrictions may be removed for specific products if a laboratory assay of treated water meets a standard as stated on the product label.
[2]Do not use treated water for irrigation in commercial nurseries or greenhouses.
[3]Refer to product label for specific restrictions.

Effectiveness of Herbicides and Triploid Grass Carp for Control of Common Aquatic Weeds in North Carolina

Weed Type	Weeds	2,4-D	bispyribac	carfentrazone	copper compounds	diquat	diquat +copper	endothall Aquathol	endothall Hydrothol	florpyrauxifen	flumioxazin	fluridone	glyphosate	imazamox	imazapyr	peroxide compounds	penoxsulam	triclopyr	triploid grass carp
Algae	Planktonic	NR	ID	NR	G	P	G	NR	P	NR	ID	NR	NR	NR	NR	G	NR	NR	NR
	Filamentous	NR	ID	NR	G	E	E	NR	E	NR	G	NR	NR	NR	NR	ID	NR	NR	P
	Chara / Nitella	NR	ID	ID	G	G	E	NR	G	NR	P	NR	NR	NR	NR	ID	NR	NR	E
Floating Plants	Azolla (mosquito fern)	NR	G	F	F	E	E	NR	NR	G	ID	E	NR	ID	NR	NR	G	NR	P
	Duckweed	P	G	G	P	G	G	NR	NR	NR	E	E	NR	NR	NR	NR	G	P	P
	Frogbit	F	ID	ID	NR	E	E	NR	NR	ID	G	NR	P	E	E	NR	ID	G	P
	Salvinia, common	NR	G	G	P	E	E	NR	NR	NR	G	E	G	E	ID	NR	ID	NR	P
	Salvinia, giant	NR	G	G	P	E	E	F	NR	NR	F	E	G	P	G	NR	E	NR	P
	Waterhyacinth	E	G	NR	NR	G	G	NR	NR	E	P	F	G	E	G	NR	E	E	P
	Watermeal	NR	NR	NR	NR	P	P	NR	NR	NR	G	G	NR	NR	NR	NR	P	NR	P
	Water lettuce	NR	G	G	NR	G	G	G	G	NR	E	NR	E	G	E	NR	E	NR	P
Emersed Plants	Alligatorweed	P	G	F	NR	NR	NR	NR	NR	G	F	F	G	G	G	NR	G	G	P
	American lotus	G	ID	NR	NR	NR	NR	NR	NR	ID	ID	G	E	F	G	NR	ID	G	P
	Cattail	F	ID	NR	NR	F	F	NR	NR	NR	P	G	E	G-E	E	NR	ID	F	P
	Creeping waterprimrose	E	ID	F	NR	NR	NR	NR	NR	G	ID	F	E	F	E	NR	G	E	P
	Floating hearts	P	ID	NR	NR	F	F	E	E	G	F	F	G	G	G	NR	F	P	P
	Fragrant waterlily	G	ID	NR	NR	NR	NR	NR	NR	E	ID	G	E	G	E	NR	ID	G	P
	Grass species	NR	ID	NR	NR	F	F	NR	NR	NR	NR	F	E	F	E	NR	ID	NR	P
	Parrotfeather	E	G	F	NR	NR	NR	NR	NR	E	F	NR	F	G	E	NR	G	E	NR
	Phragmites (Common reed)	NR	ID	NR	NR	NR	NR	NR	NR	NR	P	NR	G	F-G	E	NR	NR	F	P
	Pickerelweed	G	ID	NR	NR	NR	NR	NR	NR	E	ID	NR	F	E	E	NR	ID	G	P
	Rush	NR	ID	NR	NR	NR	NR	NR	NR	NR	ID	NR	G	ID	G	NR	ID	F	P
	Spatterdock	G	ID	NR	NR	NR	NR	NR	NR	P	ID	G	E	G	E	NR	ID	F	P
	Smartweeds	F	ID	NR	NR	F	F	NR	NR	G	ID	F	G	G	G	NR	F	G	P
	Waterpennywort	G	G	NR	NR	F	F	NR	NR	ID	G	G	E	E	E	NR	F	G	P
	Watershield	E	ID	NR	NR	F	F	NR	NR	G	ID	F	E	G	G	NR	ID	E	P
Submersed Plants	Bladderwort	P	ID	ID	NR	F	F	P	P	F	ID	E	NR	F-G	NR	NR	ID	P	E
	Cabomba	NR	ID	ID	NR	F	F	F	F	F	G	F	NR	F	NR	NR	ID	NR	F
	Coontail	G	ID	ID	NR	E	E	E	E	F	G	E	NR	NR	NR	NR	ID	G	E
	Egeria (Brazilian elodea)	NR	ID	ID	F	E	E	P	P	F	ID	E	NR	ID	NR	NR	G	NR	E
	Eurasian watermilfoil	E	G	G	NR	G	G	E	NR	E	G	E	NR	F	NR	NR	G	E	P
	Hydrilla, monoecious	NR	G	ID	F	G	E	E	E	E	G	E	NR	F	NR	NR	G	NR	E
	Naiad, brittle	NR	ID	ID	G	E	E	E	E	ID	G	E	NR	ID	NR	NR	F	NR	E
	Naiad, Southern	NR	ID	ID	G	P	G	P	P	ID	G	G	NR	ID	NR	NR	F	NR	E
	Parrotfeather	E	G	ID	NR	G	G	E	E	E	G	E	NR	F	NR	NR	G	E	F
	Pondweed species	NR	G	ID	NR	E	E	E	E	P-F	G	E	NR	G	NR	NR	G	NR	E
	Proliferating spikerush	NR	ID	ID	NR	NR	NR	NR	NR	F	P	F	NR	F	NR	NR	NR	F	E
	Variable leaf milfoil	E	ID	G	NR	E	E	E	E	E	E	G	NR	NR	NR	NR	NR	E	P

Key: NR = Not Recommended; P = Poor; G=Good ; ID = Insufficient Data; F = Fair; E = Excellent

Pond Dyes

Pond dyes may be used to prevent the growth of filamentous algae and submersed macrophyte vegetation. Pond dyes are not herbicides and do not directly kill aquatic plants. They function by blocking light penetration to the bottom of the pond. As a result, these products are most effective when applied very early in the growing season.

The use of a pond dye in aquacultural ponds usually is not recommended, as they tend to inhibit phytoplankton productivity that is needed to produce oxygen and provide food for zooplankton, which are the major food of fry and the smaller juvenile fishes. Application rates usually are about one part per million or 1 gallon per acre for a pond averaging 4 feet deep (i.e., 1 gallon per 4 acre-feet of water) for algae and most submersed weeds. For hydrilla, the rate needs to be doubled, due to its ability to grow at very low light levels. Several of the available pond dyes are registered by the USEPA for aquatic weed control. Pond dyes should not be applied to drinking water supplies or to streams or any body of water where there is any substantial outflow.

Examples of Pond Dyes	USEPA Registered
Admiral Liquid	Yes
Aquashade	Yes

Labeled Sites and Restrictions

Herbicide and Formulation	Labeled Sites	Restrictions (others may apply)[1]
2, 4-D amine (Weedar 64) 3.8 lb a.i./gal Other formulations	potable water reservoirs, farm and fish ponds, lakes, golf course water hazards, fish hatcheries	Delay the use of treated waters for irrigation and domestic purposes for 3 weeks after application unless an assay indicates that chemical water concentrates are below the minimum amount as specified on the product label. Do not treat irrigation ditches where water will be used for overhead irrigation of susceptible crops. Refer to specific product label for restrictions.
2,4-D granular (Navigate) 20 G	ponds and lakes	Do not apply to water used for irrigation, agricultural sprays, watering dairy animals, or domestic water supplies.
bispyribac (Tradewind) 80%	bayous, canals, fresh water ponds, lakes, marshes, and reservoirs	Do not irrigate until water concentrations are less than 1 ppb. Do not treat water used for crawfish production.
copper-complex (Cutrine-Plus) 0.9 lb/gal (Cutrine-Plus) 3.7 G (K-Tea) 0.8 lb/gal **copper sulfate**	potable water reservoirs, farm and fish ponds, lakes, golf course water hazards, fish hatcheries	No restrictions on use of treated water. Check tolerance of crop to copper applied in irrigation water. Trout are very susceptible to copper. Toxicity to other fish increases with decreasing hardness of water.
carfentrazone (Stingray) 1.9 lb/gal	ponds, lakes, reservoirs, marshes, wetlands, drainage ditches, canals, streams, rivers, etc.	Irrigation: Do not use treated water in commercial nurseries or greenhouses. Field crops may be irrigated after 1 day if less than 20% of surface area was treated, or after 14 days if treatment was 20% or more of surface area or until an assay indicates that chemical water concentrates are below a minimum amount as specified on the product label. Treated water may be used for turf irrigation with no restriction if less than 20% of the total water body was treated. A 14-day restriction applies for larger area treatments. Do not apply within 0.25 miles an active potable water intake (upstream only in flowing waters) or turn intake off for at least 24 hours as specified on product label. Do not drink or water livestock for 1 day if 20% or more of total surface area was treated. Applicators must be licensed or certified by the state.
diquat (Reward) 2 lb/gal	lakes, still ponds, ditches, laterals, waterways	Apply only to still water and/or public waters. Do not apply to turbid waters. Do not use treated water for irrigation of food crops, preparation of agricultural sprays, or for drinking for 5 days after application. Turf and nonfood crops may be irrigated 3 days after treatment. Do not use water for livestock for 1 day after treatment. Water use restrictions may be removed if an approved assay is conducted and water concentration is less than the maximum contaminant level as specified on product label. Refer to product label for specific PPE requirements.
dyes (Admiral Liquid) (Aquashade)	ponds and lakes with little to no outflow	Do not apply to water bodies not under direct control of user. Do not apply to water that will be used for human consumption.

Labeled Sites and Restrictions (continued)

Herbicide and Formulation	Labeled Sites	Restrictions (others may apply)[1]
endothall (Aquathol K) 4.23 lb/gal (Aquathol Super K granular) 63%	drainage canals, lakes, ponds	Consult with appropriate authorities before applying to public waters. Observe setback distance of at least 600 feet from functioning potable water intakes. Refer to specific product label for current restrictions on PPE, domestic use, irrigation, livestock use, and setback distance. Hydrothol formulations may kill fish when rates exceed 0.3 ppm.
(Hydrothol 191) 2 lb a.i./gal (Hydrothol granular) 11.2%		Check label for drinking water restrictions. Fish may be killed by Hydrothol rates exceeding 0.3 ppm. Irrigation and animal consumption restrictions of 7 to 25 days, depending on rate.
florpyrauxifen (Procellacor)	ponds, lakes, reservoirs, drainage ditches, etc.	Irrigation and livestock watering are generally restricted unless specifically allowed by product label. Do not apply to salt or brackish water. Prevent contact to or drift on sensitive species. Do not use with organosilicone surfactants.
flumioxazin (Clipper) 51%	bayous, canals, fresh water ponds, lakes, marshes, and reservoirs	Do not irrigate from treated water for at least 5 days. Do not treat water used for crawfish production.
fluridone (Sonar 4 AS or SRP)	lakes, ponds, canals	Treated ponds may not be used for irrigation for 7 to 30 days. See label for irrigation precautions.[1]
glyphosate (AquaMaster) 5.4 lb a.i./gal (AquaNeat) 5.4 lb a.i./gal (Rodeo) 5.4 lb a.i./gal (Touchdown Pro) 3 lb a.e./gal Other formulations	all bodies of fresh water and all types of aquatic sites	Do not apply within 0.5 mile of an active potable water intake (upstream only in flowing waters) or turn intake off for at least 48 hours as specified on product label. Refer to specific product label for restrictions.
imazamox (Clearcast)	in and around aquatic and noncropland sites	Irrigation: Do not apply to water to be used for irrigation of greenhouse or nursery plants. Do not irrigate from still or quiescent water bodies within 24 hours of application. Do not irrigate if concentrations exceed 50 ppb.
imazapyr (Habitat)	in and around standing and flowing waters, including estuarine and marine sites	Irrigation: Do not use treated water for 120 days following application or until an assay indicates that chemical water concentrations are below a minimum amount as specified on the product label. Do not apply within 0.5 mile of an active potable water intake (upstream only in flowing waters) or turn intake off for at least 48 hours as specified on product label. Do not apply to fast-moving waters. Do not apply to irrigation ditches or canals within 1 mile of an active irrigation water intake unless the irrigation restrictions can be observed. Applicators must be licensed or certified by the state.
penoxsulam (Galleon)	in and around quiescent water bodies and exposed sediments of de-watered areas	Do not apply to flowing water. Irrigation: Do not apply to water to be used for irrigation of greenhouse or nursery plants. Do not irrigate established food crops, other than rice, if concentrations exceed 1 ppb. Do not irrigate established rice if concentrations in treated water exceed 30 ppb. No restrictions on use of treated water for turf irrigation, if concentrations are less than 30 ppb. Consult SePRO for other situations/commodities.
sodium carbonate peroxyhydrate (GreenClean Pro) (PAK 27)	ponds, lakes, lagoons, canals, ditches, etc.	Do not apply to treated, finished drinking water reservoirs.
triclopyr (Renovate 3) 3 lb/gal (Renovate OTF) 10 G	quiescent and slow-moving waters; non-irrigation canals	Irrigation: Do not use treated water for 120 days following application or until treated water has a non-detectable triclopyr level by an assay as specified on the product label. No restriction on irrigation of established grass. Applications around potable water intakes must observe minimum setback distances and/or minimum water concentrations as specified on the product label. Do not apply directly to or allow to come in direct contact with grapes, tobacco, vegetable crops, flowers, and other desirable broadleaf plants. Do not apply to estuarine sites; do not apply directly to un-impounded rivers or streams; and do not apply to irrigation ditches or canals. Do not allow lactating dairy animals to graze treated areas until the next growing season after application unless spot-treatment was applied to less than 10% of total grazable area. Animals for slaughter must be removed from the treated area for at least 3 days. Do not treat more than ½ of water body in a single operation; wait 10 to 14 days for next treatment.

[1]Water use restrictions for irrigation vary with formulation. See label for precautions. A 30-day restriction may be insufficient if applied to small ponds intended for irrigation of very sensitive crops, such as tobacco, tomatoes, or peppers.

Integrated Pest Management: The Sensible Approach to Turf Care

Many pest problems can cause your turf to look bad — diseases, weeds, insects, and animals. Some people have all of these problems. Is a pesticide the proper solution? Or is it better to make changes in cultural practices? Both methods, and some others as well, may be needed. The balanced use of all available methods is called Integrated Pest Management (IPM).

The idea is simple. It involves using all available prevention and control methods to keep pests from reaching damaging levels. The goal is to produce good turf and minimize the influence of pesticides on people, the environment, and turf.

IPM methods include:

1. use of best-adapted grasses;

2. proper use of cultural practices, such as watering, mowing, and fertilization; and

3. proper selection and use of pesticides when necessary.

Early detection and prevention will minimize pest damage, saving time, effort, and money. Should a problem occur, determine the cause or causes, then choose the safest, most effective control or controls available.

When chemical control is necessary, select the proper pesticide, follow label directions, and apply when the pest is most susceptible. Treat only those areas in need. Regard pesticides as only one of many tools available for turf care.

More information about IPM, pest identification, turf care, and proper use of pesticides, is available on the Web at www.turffiles.ncsu.edu. A Cooperative Extension agent in your county may also be of assistance.

Misuse of Pesticides

It is a violation of the law to use any pesticide in a manner not permitted by its labeling.

As a protection from violating the law, never apply any pesticide in a manner or for a purpose other than as instructed on the label or in labeling accompanying the pesticide product. Don't ignore the instructions for use of protective clothing and devices and for storage and disposal of pesticide wastes, including containers. All recommendations for pesticide use included in this publication were legal as of November 2016, but the status of registration and use patterns is subject to change by actions of state and federal regulatory agencies.

Help Make Extension Better—Contribute to Our Publications.

NC State Extension helps to strengthen North Carolina families and communities every day through our mission and outreach programs. Our publications and communications enhance Extension's statewide, regional, and county programmatic efforts. Your contribution will support the production of these publications, help empower people, and provide solutions.

Make a secure gift online: **go.ncsu.edu/ExtPublications**

☐ A check to support Extension publications for the total amount of $ _____ is enclosed.
Please make checks payable to the North Carolina Agricultural Foundation, Inc. and note Account #011893.

☐ Please contact me about making my gift in bank drafts or appreciated stocks.

NAME _____

ADDRESS _____

CITY _____ STATE _____ ZIP _____

PHONE _____ EMAIL _____

Fundraising efforts for Extension Publications/Communications operate under the auspices of the North Carolina Agricultural Foundation, Inc., a 501(c)3 non-profit (Tax ID# 56-6049304). You will receive an official receipt for your tax-deductible donation.

Please contact cals_advancement_business@ncsu.edu or 919.515.2000 with questions regarding donations.

Mail checks or contact information to

CALS Advancement
NC State University
Campus Box 7645
Raleigh, NC 27695-7645